Spectral Mapping Theorems

Robin Harte

Spectral Mapping Theorems

A Bluffer's Guide

Second Edition

Springer

Robin Harte
School of Mathematics
Trinity College
Dublin, Ireland

ISBN 978-3-031-13916-1   ISBN 978-3-031-13917-8   (eBook)
https://doi.org/10.1007/978-3-031-13917-8

Mathematics Subject Classification: 47-01, 47-02

This Springer imprint is published by the registered company Springer Nature Switzerland AG
The registered company address is: Gewerbestrasse 11, 6330 Cham, Switzerland

Dedicated to the memory of Donal O'Donovan and Richard Timoney.

Preface to the Second Edition

In this second edition of the "bluffer's guide", we have corrected as many misprints and typos as we could identify in the first and made some additions to the bibliography. We have also taken the opportunity to sharpen the presentation of the spectral mapping theorem for the several variable left spectrum: we have noticed that the platform from which the proof is launched is an old idea from commutative ring theory, the "residual quotient" of ideals. We have at the same time revised our treatment of Taylor invertibility, where we have to deal the exactness of an associated "Koszul complex", which can be realized as a vector-valued differential form; we here as an alternative present a rather simple inductive version. For the several variable Taylor spectral mapping theorem, we observe that the (splitting) exactness depends on whether or not a certain sum of a left and a right ideal contains the identity; this now needs a "two-sided" version of the residual quotient, which turns up as spin-off from the solution of a problem in approximation theory.

We have in this second edition added new material on the block structure of "Operator Matrices" and the consequences of the "spectral disjointness" of their diagonal entries. We have also discussed Wawrzynczyk's extension of the spectral mapping theorem from Banach to "Waelbroeck" algebras, and as a little flourish at the end, we notice that elementary number theory can be made to look like spectral theory.

We have maintained the six chapter structure of the first edition, comprising three introductory chapters on algebra, topology and topological algebra, followed by three on spectral theory: of one, several and "many" variables.

Throughout this work, Michael Mackey, of NUI Dublin, and Carlos Hernandez, of UNAM Mexico, have been crucial sounding boards, ready also at the drop of a hat, to pick up missing dollar signs and other typos. From Dublin City University, Charlie Daly has kept my little computer afloat, and Jane Horgan gifted me the Latex preamble which she created for the second edition of her ground-breaking Wiley tome "Probability with R" [2020] and has continued to track down elusive latex errors, while Mehmet Aygünes has promised to translate the whole thing into Turkish.

Dublin, Ireland   Robin Harte

Preface to the First Edition

Suppose $ab = ba \in A$, a commuting pair of Banach algebra elements $(a, b) \in A^2$; then

$$\sigma(a + b) \subseteq \sigma(a) + \sigma(b) \; ; \; \sigma(ab) \subseteq \sigma(a)\sigma(b) \; :$$

the spectrum of the sum and the product are subsets of the sum and the product of the spectra. One way to prove this is via *Gelfand's theorem*: find a "maximal abelian" subalgebra $D \subseteq A$ for which $\{a, b\} \subseteq D$, and argue

$$\sigma_A(a + b) = \{\varphi(a + b) : \varphi \in \sigma(D)\} \subseteq \{\psi(a) + \theta(b) : \{\psi, \theta\} \subseteq \sigma(D)\}$$

and similarly

$$\sigma_A(ab) = \{\varphi(ab) : \varphi \in \sigma(D)\} \subseteq \{\psi(a)\theta(b) : \{\psi, \theta\} \subseteq \sigma(D)\} \; .$$

Here $\sigma(D) \subseteq D^*$ is the "maximal ideal space" of the commutative Banach algebra D and we need to know that there is implication

$$c \in D \Longrightarrow \sigma_A(c) = \sigma_D(c) = \{\varphi(c) : \varphi \in \sigma(D)\} \; .$$

Much sweeter would be a *joint spectrum* argument, enabling us to write

$$\sigma(a + b) = \{\lambda + \mu : (\lambda, \mu) \in \sigma(a, b)\} \; ; \; \sigma(ab) = \{\lambda\mu : (\lambda, \mu) \in \sigma(a, b)\}$$

with

$$\sigma(a, b) \subseteq \sigma(a) \times \sigma(b) \; .$$

That, in a nutshell, is what these notes are all about.

In more detail, we set out to describe the *spectral mapping theorem* in one, "several" and "many" variables. As background we need to introduce the basic algebraic systems, including *semigroups*, *rings* and *linear algebras*.

In a sense, abstract algebra is the story of a continuing negotiation between the "invertible" and the "singular". The invertibles, followers of "one", tread lightly in the garden, doing nothing that cannot be undone; the singulars however follow "zero", and have an altogether heavier tread. The garden where all these games are to be played will be something called a *Banach algebra*, which, like an urban environment, is rich in interlocking structures. It may therefore be a good idea to step back a little to a much more primitive rural setting. Algebra at its most basic is played out in a *semigroup*, a system with a "multiplication"; however most of the interesting semigroups are *rings*, which are semigroups twice over, with multiplication and "addition" tied together by a *distributive law*. In turn most of the interesting rings are *linear algebras*, submitting to real or complex number multiplication. Basic algebra can also be extended by taking *limits*, whose study is sometimes called "point set topology": in a "Banach algebra" this topology is generated by a norm and a metric structure, which has to be "complete". It will be a bonus that almost everything we say about semigroups, or rings, remains valid in abstract *categories*, or *additive categories*.

Together with algebra we need therefore to discuss *topology*, the abstract theory of limits. Now *topological algebra* is about algebraic systems which also have topology, in such a way that the fundamental algebraic operations are continuous. Our main focus is to specialise to complex Banach algebras, where we find that the spectrum is truly well-behaved: nonempty, compact, and subject to the spectral mapping theorem.

Even in semigroups and rings the invertible and the singular are locked into a minuet: usually products of invertibles are again invertible while often sums of invertible and singular are invertible. Homomorphisms $T : A \rightarrow B$ between semigroups or rings send invertibles to invertibles, $T(A^{-1}) \subseteq B^{-1}$, while there are two ways for a homomorphism to behave well. One, relating to invertibility, would be the *Gelfand property* $T^{-1}B^{-1} \subseteq A^{-1}$, while the other, relating to singularity, would be a *Riesz property*, which says that for example $A^{-1} + T^{-1}(0) \subseteq A$ is "almost" in A^{-1}.

In linear algebras we meet the idea of the *spectrum*, with which we can draw real or complex pictures following the progress of the invertible/singular debate; this then harnesses complex analysis to the theory: rather than being simply an elegant commentary on events, the complex pictures to a considerable extent drive the action.

Spectral theory therefore is dedicated to the theory of invertibility: the spectrum of an algebra element $a \in A$ simply collects those scalars λ which give perturbations $a - \lambda$ which fail to be invertible. Generally various kinds of "non singularity" are necessary conditions for invertibility: each of them generates their own subset of the spectrum.

All this is for single elements of a Banach algebra; the extension to n tuples of elements gives spectra consisting of n tuples of complex numbers: the fundamental theorem says that for *commuting* tuples of elements we still have non empty spectrum and the *spectral mapping theorem for polynomials*. Our "joint spectrum" could be the union of a "left" and a "right" spectrum, but this turns out to be deficient in the sense that it does not in general support a *functional calculus*: that needs the more sophisticated ideas, due to Joseph Taylor, based on exactness.

There are canonical extensions, based on compactness, from finite tuples to infinite systems: what we find interesting here is what happens when there is additional algebraic or topological structure on the indexing material.

In *several variables* there are various ways in which an n-tuple $a = (a_1, a_2, \ldots, a_n) \in A^n$ can be considered to be "invertible", thus generating a "spectrum" $\sigma(a) \subseteq \mathbb{C}^n$. The extension to *many variables* looks at systems $a \in A^X$ indexed by arbitrary sets X which now possibly carry algebraic or topological structure, capable of being respected by the system a; here the story is that whenever this happens it continues to happen to $\lambda \in \sigma(a) \subseteq \mathbb{C}^X$. However the true "multi-variate" extension of the concept of invertibility is the idea of *exactness*.

One version of the "joint spectrum" of a commuting tuple, and its spectral mapping theorem, can indeed be derived from "maximal abelian" subalgebras, and Gelfand's theorem for commutative Banach algebras; conversely our spectral mapping theorem for several variables, and its many variable extension, give an alternative proof of that same Gelfand theorem.

The first appearance of the definition of the left and the right spectrum of an n tuple seems to be in the numerical range notes [29] of Bonsall and Duncan. The proof of the spectral mapping theorem for commuting finite tuples was first given, for bounded operators on Hilbert space, by John Bunce [35]: the argument involved C* algebras and states. An equivalent argument, without any definition of joint spectrum, is part of the Graham Allen paper [4] about holomorphic left inverses. The spectral mapping theorem for the "Taylor spectrum" is in the first [297] of the two ground breaking papers of Joseph Taylor: the second of these goes on to establish the "functional calculus", which successfully defines $f(T)$ when $f : U \to \mathbb{C}$ is holomorphic on an open neighbourhood U of the "Taylor spectrum" of a commuting tuple $T = (T_1, T_2, \ldots, T_n)$ of Banach space operators. As Vladimir Müller has demonstrated this is significantly easier to describe for functions holomorphic on the sometimes larger "Taylor split spectrum". The split spectrum has a natural extension to tuples of Banach algebra elements; we do not here offer any extension to Banach algebra elements of the original Taylor spectrum of operators. Discussion of an abstract idea of spectrum can be traced back to Wieslaw Zelazko [321]; Vladimir Müller [259] has cast this in the framework of his concept of a "regularity". In one variable, this is a refinement of a more primitive idea contained in our discussion of "Kato invertibility" and non singularity [145, 146, 148]. Lucien Waelbroeck [304, 305], and later Vladimir Kisil [211, 212], have viewed a "spectrum" as the support of a more primitive "functional calculus".

The whole point of the additional subtleties of the Taylor spectrum and split spectrum is the functional calculus [63 66, 199]. Generations of mathematicians have been trying to capture and regulate the intuition of the engineer Heaviside, and for many the test of a "spectrum" will always be whether it supports a "functional calculus". In these notes however we stop short with the spectral mapping theorem.

In telling our story we have divided the narrative into six chapters: the first three, in introductory mode, start with pure algebra and progress, through topology, to "topological algebra", while the remaining three are officially spectral theory, progressing from one through several to "many" variables. What we have collected together as pure algebra sometimes, in other versions of the story, appears on the far side of a build up of topological or metric structures. Similarly what we have collected together as topology is sometimes introduced piecemeal in the middle of other argument. We have also tried, in a sense, to present topology as a kind of algebra: for example the relationship between topological boundary and connected hull can be given this flavour.

It is clear that there is significant overlap between these notes and our earlier volume, "Invertibility and Singularity" [132]. The treatment here is quite different however: in our earlier book we give complete proofs of such things as the Hahn-Banach, Open Mapping and Liouville theorems, here omitted; we also devote attention to what happens for bounded operators between incomplete normed spaces. We believe that strategy offered valuable insight into the open mapping theorem: between incomplete spaces open, almost open and onto are in general different kinds of linear operator, which all coalesce when the spaces are complete. Spectral and Fredholm theory can also be discussed in an incomplete environment; with hindsight however the extra effort proves distracting, and in the present notes we abandon this discussion, looking instead sometimes at the analagous discussion in the purely linear environment. While there is therefore a good deal of material in the earlier volume which has not been reproduced here, there is also, thanks to the passage of time, a great deal of material here which was not available to the earlier work.

There is also overlap between both this and our earlier work with the book, "Spectral theory of linear operators" [259], of Vladimir Müller. There the spectral theory of one and several variables is systematically expounded through the medium of the Müller concept of "regularity". The single variable version of this concept, and its more primitive "non commutative" version, can be traced back to our own journal papers [145, 146, 148] of 1992, 1993 and 1996.

Contents

Algebra

1

The language of spectral theory is algebra, so we should begin with some kind of dictionary.

1.1 Semigroups

The basic algebraic structure for us will be a semigroup, , in the sense of a nonempty set A supporting a binary operation

$$* : (x, y) \mapsto xy = x * y \ (A \times A \to A) \tag{1.1.1}$$

which satisfies the *associative law*:

$$(xy)z = (x * y) * z = x * (y * z) = (xy)z \ (x, y, z \in A) . \tag{1.1.2}$$

On account of (1.1.2) we feel free to write

$$(xy)z = xyz \ (x, y, z \in A) . \tag{1.1.3}$$

An element $1 \in A$ will be called an *identity* if

$$1x = 1 * x = x = x * 1 = x1 \ (x \in A) : \tag{1.1.4}$$

the very first exercise for the reader is to see that there can never be more than one identity in a semigroup A. Imperceptibly we shall modify the meaning of the word "semigroup" to add the assumption that there is an identity $1 \in A$. At the opposite extreme to an identity is a *zero* , in the sense of an element $0 \in A$ for which

$$0x = 0 * x = 0 = x * 0 = x0 \ (x \in A) . \tag{1.1.5}$$

© The Author(s), under exclusive license to Springer Nature Switzerland AG 2023
R. Harte, *Spectral Mapping Theorems*,
https://doi.org/10.1007/978-3-031-13917-8_1

The second exercise for the reader is to see that there can never be more than one zero in a semigroup. For a third exercise, the reader is invited to enquire about the possibility that an identity could also be a zero; this can happen, but only in what we will agree is a rather trivial situation: the *singleton*

$$\mathbb{O} = \{0\} \ with \ 0 * 0 = 0 \,. \tag{1.1.6}$$

The singleton can only carry one possible binary operation, and is indeed a bona fide semigroup.

The reader is invited to reflect that all the familiar number systems $\mathbb{N}, \mathbb{Z}, \mathbb{Q}, \mathbb{R}, \mathbb{C}$ are semigroups twice over, relative to *addition* and to *multiplication*. The set X^X of all mappings on a set X is also a semigroup, relative to *composition*. The set 2^X of all subsets of a set X is a semigroup twice over: relative to *union*, and to *intersection*. If $X = A$ is itself a semigroup then so in a third sense is the set 2^X of its subsets, if we define

$$KH = K * H = \{x * y : x \in K, y \in H\} \,; \tag{1.1.7}$$

conversely the left and right residual quotients of H by K are given by

$$K^{-1}H = \{x \in A : Kx \subseteq H\} \,, \ HK^{-1} = \{x \in A : xK \subseteq H\} \,. \tag{1.1.8}$$

Evidently (1.1.8) is monotonically increasing in H and decreasing in K:

$$(H' \subseteq H \ \& \ K \subseteq K') \Longrightarrow K'^{-1}H' \subseteq K^{-1}H \,, \tag{1.1.9}$$

$$KH \subseteq H \Longleftrightarrow H \subseteq K^{-1}H \,, \tag{1.1.10}$$

and

$$H \subseteq K \Longrightarrow (K^{-1}H)(K^{-1}H) \subseteq K^{-1}H \,. \tag{1.1.11}$$

When there is an identity $1 \in A$ there is implication

$$K \subseteq H \Longrightarrow 1 \in K^{-1}H \,; \tag{1.1.12}$$

$$1 \in K \Longrightarrow K^{-1}H \subseteq H \,. \tag{1.1.13}$$

If we were feeling subversive, we might remark that much of what we have to say about semigroups-with-identity can be extended to *abstract categories*. Technically in fact an abstract category is nothing more than a very large semigroup—too large to be a set—with partially defined binary operation and many different identities. As in for example the category of sets and mappings, every element, or *morphism*, comes with its *departure object* and its *arrival object,* conceived of as identities rather than the sets they live on.

1.2 Invertibility

If A is a semigroup, with identity 1, then a left inverse for an element $x \in A$ will be an element $x' \in A$ for which

$$x'x = 1 , \tag{1.2.1}$$

and a right inverse an element $x'' \in A$ for which

$$xx'' = 1 . \tag{1.2.2}$$

If we write

$$A^{-1}_{left} = \{x \in A : 1 \in Ax\} , \ A^{-1}_{right} = \{x \in A : 1 \in xA\} \tag{1.2.3}$$

for the sets of left and of right invertible elements then they are also semigroups in their own right; indeed for arbitrary x, y in A there is implication

$$\{x, y\} \subseteq A^{-1}_{left} \Longrightarrow xy \in A^{-1}_{left} \Longrightarrow y \in A^{-1}_{left} \tag{1.2.4}$$

and

$$\{x, y\} \subseteq A^{-1}_{right} \Longrightarrow xy \in A^{-1}_{right} \Longrightarrow x \in A^{-1}_{right} . \tag{1.2.5}$$

Generally a left invertible can have more than one left inverse, and a right invertible more than one right inverse: but as soon as a left invertible also has a right inverse then there can only be one. The reader may like to check that there is implication

$$x'x = 1 = xx'' \Longrightarrow x' = x'' . \tag{1.2.6}$$

We shall write

$$A^{-1} = A^{-1}_{left} \cap A^{-1}_{right} \tag{1.2.7}$$

for the crucially important set of *invertibles* , and as a matter of notation

$$x'x = 1 = xx'' \Longrightarrow x' = x'' = x^{-1} \tag{1.2.8}$$

will be the *inverse* of $x \in A$. Note particularly the *reverse order law*: if $x, y \in A^{-1}$ then

$$1^{-1} = 1; \; (x^{-1})^{-1} = x \,; \; (xy)^{-1} = y^{-1}x^{-1} \,. \tag{1.2.9}$$

In partial converse to (1.2.4) and (1.2.5),

$$\left(xy \in A_{left}^{-1} \,, \; y \in A_{right}^{-1}\right) \Longrightarrow x \in A_{left}^{-1} \tag{1.2.10}$$

and

$$\left(xy \in A_{right}^{-1} \,, \; x \in A_{left}^{-1}\right) \Longrightarrow y \in A_{right}^{-1} \,. \tag{1.2.11}$$

As a common generalization of left and right inverses, a *generalized inverse* for $x \in A$ is $y \in A$ for which

$$x = xyx \,. \tag{1.2.12}$$

We shall write

$$A^{\cap} = \{x \in A : x \in xAx\} \tag{1.2.13}$$

for the set of such *relatively regular* elements. Evidently (exercise)

$$A^{-1} \subseteq A_{left}^{-1} \cup A_{right}^{-1} \subseteq A^{\cap} \,. \tag{1.2.14}$$

If $y \in A$ satisfies (1.2.12) then so does $y' = yxy$, in which case also $y' = y'xy'$: thus we can always arrange that the relationship between an element and its generalized inverse is symmetric. On the other hand it is also possible that a generalized inverse is itself invertible: we shall write

$$A^{\cup} = \{x \in A : x \in xA^{-1}x\} \tag{1.2.15}$$

for the set of these *invertibly regular* elements. The reader may like to verify

$$A^{\cup} \cap A_{left}^{-1} = A^{\cup} \cap A_{right}^{-1} = A^{-1} \,. \tag{1.2.16}$$

There is no analogue of either part of (1.2.4) or (1.2.5) for generalized inverses.

Notice how two-sided invertibility for elements a, b and their product ba, form a *democratic consensus*; by (1.2.10) and (1.2.11), of the three conditions

$$a \in A^{-1} \,, \; b \in A^{-1} \,, \; ba \in A^{-1} \,: \tag{1.2.17}$$

each is a consequence of the other two.

1.3 Zero Divisors

If the semigroup A has a zero element 0 then it will also have *zero divisors* , which come in two flavours, left and right. This is particularly the case when A is a ring, with 0 the additive identity. If we define mappings $L_x : A \to A$ and $R_x : A \to A$ by setting

$$L_x(y) = xy , \quad R_x(y) = yx \ (x, y \in A) , \tag{1.3.1}$$

then we can write

$$A^o_{left} = \{x \in A : L_x^{-1}(0) = \{0\}\} , \quad A^o_{right} = \{x \in A : R_x^{-1}(0) = \{0\}\} ; \tag{1.3.2}$$

we shall sometimes call these elements monomorphisms and epimorphisms . Now $x \in A$ will be called a *left zero divisor* if it is not in A^o_{left}, and a *right zero divisor* if it is not in A^o_{right}. Evidently (exercise!)

$$A^{-1}_{left} \subseteq A^o_{left} , \quad A^{-1}_{right} \subseteq A^o_{right} , \tag{1.3.3}$$

and also

$$A^{-1} = A^{-1}_{left} \cap A^o_{right} = A^{-1}_{right} \cap A^o_{left} . \tag{1.3.4}$$

Unless $A = \{0\}$ is a singleton the zero 0 is both a left and a right zero divisor. The reader may like to check that none of the number systems have any zero divisors other than zero. Analogous to (1.2.4) and (1.2.5) there is implication, for arbitrary x, y in A,

$$\{x, y\} \subseteq A^o_{left} \implies xy \in A^o_{left} \implies y \in A^o_{left} \tag{1.3.5}$$

and

$$\{x, y\} \subseteq A^o_{right} \implies xy \in A^o_{right} \implies x \in A^o_{right} . \tag{1.3.6}$$

As an improvement to (1.2.10) and (1.2.11), and a converse to (1.3.5) and (1.3.6),

$$\left(xy \in A^{-1}_{left} , \ y \in A^o_{right}\right) \implies x \in A^{-1}_{left} \tag{1.3.7}$$

and

$$\left(xy \in A^{-1}_{right} , \ y \in A^o_{left}\right) \implies x \in A^{-1}_{right} , \tag{1.3.8}$$

and

$$\left(xy \in A^o_{left} , \ y \in A^o_{right}\right) \implies x \in A^o_{left} \tag{1.3.9}$$

and

$$\left(xy \in A_{right}^o \,,\, y \in A_{left}^o\right) \Longrightarrow x \in A_{right}^o \,. \tag{1.3.10}$$

Also

$$A^\cap \cap A_{left}^o = A_{left}^{-1} \,,\, A^\cap \cap A_{right}^o = A_{right}^{-1} \,, \tag{1.3.11}$$

and, improving (1.2.7),

$$A^\cup \cap A_{left}^o = A^\cup \cap A_{right}^o = A^{-1} \,. \tag{1.3.12}$$

For example the *nilpotents* $x \in A$ for which

$$0 \in \{x^n : n \in \mathbb{N}\} \tag{1.3.13}$$

are both left and right zero divisors.

Notice how the relatively regular elements of (1.2.13) are characterized by the multiplication operators of (1.3.1):

$$a \in A^\cap \Longleftrightarrow a \in L_a R_a(A) \,; \tag{1.3.14}$$

$$a \in A^\cup \Longleftrightarrow a \in L_a R_a(A^{-1}) \,. \tag{1.3.15}$$

1.4   Commutivity

We say that $x \in A$ and $y \in A$ commute if

$$yx = xy \,, \tag{1.4.1}$$

and if (1.4.1) holds for every pair $x, y \in A$ then the semigroup A is said to be *commutative*, or "abelian". Generally if $K \subseteq A$ we write, for the *commutant* of K in A, comm(K)

$$\mathrm{comm}(K) \equiv \mathrm{comm}_A(K) = \{x \in A : \forall y \in K : yx = xy\} \,. \tag{1.4.2}$$

For example the commutant of $x \in A$ is comm(x) = comm(K) with $K = \{x\}$, and the *double commutant* comm$^2(x)$ is the commutant of the commutant:

$$\mathrm{comm}^2(x) = \mathrm{comm\,comm}(x) \,. \tag{1.4.3}$$

Commutants are always sub semigroups, and the reader might like to verify that

$$x \in A^{-1} \Longrightarrow x^{-1} \in \mathrm{comm}^2(x) \,. \tag{1.4.4}$$

Obviously any left inverse which commutes must be a, therefore the, two-sided inverse. We shall also have use for a "commuting" variant of the set product (1.1.7), writing for $K, H \subseteq A$

$$K *_{comm} H = \{x * y : x \in K , y \in H , x * y = y * x\} . \tag{1.4.5}$$

A generalized inverse which commutes is the next best thing to an inverse: we shall call $x \in A$ *simply polar* , or *group invertible*, if there is $y \in \mathrm{comm}(x)$ satisfying (1.2.12), and declare the *group inverse* of such $x \in A$ to be $y \in A$ for which

$$x = xyx ; \quad xy = yx ; \quad y = yxy . \tag{1.4.6}$$

It turns out that such y is uniquely determined, and double commutes with x. Necessary and sufficient for $x \in A$ to have a group inverse is that, writing of course $xx = x^2$,

$$x \in (x^2 A) \cap (A x^2) : \tag{1.4.7}$$

note that

$$x^2 u = x = v x^2 \Longrightarrow (xu = vx \text{ and } xux = xvx)$$

and hence $y = vxu$ satisfies (1.2.12) and commutes with x.

Invertible elements are group invertible, and group invertible elements are invertibly regular in the sense (1.2.15). *Idempotents*

$$x = x^2 \tag{1.4.8}$$

are also group invertible; notice generally that if $x = xyx$ is relatively regular then both products xy and yx are idempotent . The reader may like to verify that the group invertibles coincide with the commuting products of invertibles and idempotents. We shall also write

$$\mathrm{SP}(A) = \{x \in A : x \in (x^2 A) \cap (A x^2)\} \tag{1.4.9}$$

for the group invertibles in A, and refer to them as *simply polar*: evidently

$$\mathrm{SP}(A) \subseteq A^{\cup} \tag{1.4.10}$$

and hence

$$\mathrm{SP}(A) \cap A^{-1}_{left} = A^{-1} = \mathrm{SP}(A) \cap A^{-1}_{right} . \tag{1.4.11}$$

We shall also write, for $K \subseteq A$,

$$\text{comm}^{-1}(K) = A^{-1} \cap \text{comm}(K) . \tag{1.4.12}$$

We sometimes refer to $\text{comm}(A)$ as the *Centre* of A.

Notice that products of commuting idempotents are again idempotent: if

$$(p, q) = (p^2, q^2) \in A \times A \tag{1.4.13}$$

then if also $pq = qp$ then we have

$$(pq)^2 = pq . \tag{1.4.14}$$

Commutivity $pq = qp$ is sufficient but not necessary for (1.4.14): for example, with (1.4.13),

$$(1 - q)(1 - p) = 0 \tag{1.4.15}$$

is also sufficient for (1.4.14). Notice, with only (1.4.13), that, with

$$u = qp + (1 - q)(1 - p) , \ v = pq + (1 - p)(1 - q) , \tag{1.4.16}$$

we have

$$vu = 1 - (p - q)^2 = uv . \tag{1.4.17}$$

The invertibility of sums, products and differences of idempotents are related: again with (1.4.13), there is equivalence

$$p - q \in A^{-1} \Longleftrightarrow \{p + q, 1 - pq\} \subseteq A^{-1} , \tag{1.4.18}$$

provided $1 + 1 \in A^{-1}$.

1.5   Homomorphisms

Mathematical systems, often consisting of sets with additional structure, always come with "homomorphisms", and semigroups are no exception. A *homomorphism* $T : A \to B$ of semigroups is a mapping for which

$$T(xy) = (Tx)(Ty) \ (x, y \in A) ; \tag{1.5.1}$$

if there are identities $e = 1 \in A$ and $f = 1 \in B$ it will be understood that also

$$T(e) = f . \tag{1.5.2}$$

The reader is invited to confirm that it follows

$$T(A^{-1}) \subseteq B^{-1} , \tag{1.5.3}$$

and more generally

$$T(A_{left}^{-1}) \subseteq B_{left}^{-1} , \quad T(A_{right}^{-1}) \subseteq B_{right}^{-1} . \tag{1.5.4}$$

Homomorphisms bring into play a sort of "Fredholm theory": (1.5.3) translates to inclusion

$$A^{-1} \subseteq T^{-1} B^{-1} \subseteq A . \tag{1.5.5}$$

When there is equality in (1.5.5) we shall say that the homomorphism $T : A \to B$ has the *Gelfand property*

$$T^{-1}(B^{-1}) \subseteq A^{-1} \subseteq A : \tag{1.5.6}$$

In words, "Fredholm implies invertible". If for example $K \subseteq A$ is a subset with the property

$$1 \in KK \subseteq K \tag{1.5.7}$$

then it can be thought of as a semigroup in its own right; now the *natural injection*

$$J_K : x \mapsto x \ (K \to A) \tag{1.5.8}$$

has the status of a homomorphism. If A and B are semigroups then so is their *direct sum* $A \oplus B$, derived from the cartesian product $A \times B$ by setting

$$(x', y')(x, y) = (x'x, y'y) \ (x, x' \in A, y, y' \in B) . \tag{1.5.9}$$

If X is a set then the set A^X of mappings from X to A is a semigroup, with

$$(xy)(t) = x(t)y(t) \ (t \in X ; \ x, y \in A^X) . \tag{1.5.10}$$

Each $t \in X$ induces a homomorphism

$$x \mapsto x(t) : A^X \to A ; \tag{1.5.11}$$

similarly there are two homomorphisms $A \oplus B \to A$ and $A \oplus B \to B$ given by

$$(a, b) \mapsto a ; \ (a, b) \mapsto b . \tag{1.5.12}$$

The left regular representation $T = L$

$$x \mapsto L_x : A \to A^A \qquad (1.5.13)$$

of (1.3.1) has the Gelfand property (1.5.6):

$$L_{xy} = L_x \circ L_y \ (x, y \in A) . \qquad (1.5.14)$$

The reader is invited to reflect on the status of the mapping $x \mapsto R_x$. If

$$A = K = \text{comm}(H) \subseteq B \qquad (1.5.15)$$

then the natural injection $T : A \to B$ of (1.5.8) also has, by (1.4.4), the Gelfand property.

If $T : A \to B$ is a homomorphism then

$$T(A^\cap) \subseteq B^\cap \subseteq B \qquad (1.5.16)$$

and hence

$$A^\cap \subseteq T^{-1}(B^\cap) \subseteq A ; \qquad (1.5.17)$$

if there is equality here we shall say that the homomorphism T has *generalized permanance* . We shall also refer to the Gelfand property as "spectral permanence". When $T : A \to B$ is a homomorphism there is also inclusion

$$T \ SP(A) \subseteq SP(B) \subseteq B , \qquad (1.5.18)$$

and

$$SP(A) \subseteq T^{-1}SP(B) \subseteq A . \qquad (1.5.19)$$

equality in (1.5.19) will be called "simple permanence" .

The "alter ego" of a homomorphism is an *antihomomorphism* , $T : A \to B$ for which (1.5.1) is replaced by

$$T(xy) = (Ty)(Tx) \ (x, y \in A). \qquad (1.5.20)$$

An immediate example is the *right regular representation*

$$x \mapsto R_x$$

of (1.3.1). A special kind of antihomomorphism is an *involution* $x \mapsto x^* : A \to A$ for which

$$(xy)^* = y^*x^* \; ; \; (x^*)^* = x \; ; \; 1^* = 1 \; . \tag{1.5.21}$$

In the presence of an involution we can distinguish a special kind of generalized inverse for $a \in A$: $b \in A$ for which

$$a = aba \; ; \; b = bab \; ; \; (ba) = (ba)^* \; ; \; ab = (ab)^* \; . \tag{1.5.22}$$

If in particular there is *cancellation* , in the sense of implication, for arbitrary $a, x \in A$,

$$a^*ax = 0 \Longrightarrow ax = 0 \; , \tag{1.5.23}$$

then the conditions (1.5.22) uniquely determine the *Moore-Penrose inverse*

$$b = a^\dagger \tag{1.5.24}$$

of $a \in A$. Further, there is inclusion

$$A^\dagger \subseteq SP^*(A) \subseteq A^\cap \; , \tag{1.5.25}$$

where

$$A^\dagger = \{a \in A : \exists \, a^\dagger \in A\} \tag{1.5.26}$$

and we write $SP^*(A)$ for the "star polar" elements of A,

$$SP^*(A) = \{a \in A : a^*a \in A^\cap\} \; . \tag{1.5.27}$$

Of course, if $x \mapsto x^*$ is an involution on A, then there is implication, for $a \in A$,

$$a \in A_{left}^{-1} \Longleftrightarrow a^* \in A_{right}^{-1} \; , \tag{1.5.28}$$

and hence

$$a = a^* \in A_{left}^{-1} \cup A_{right}^{-1} \Longrightarrow a \in A^{-1} \; . \tag{1.5.29}$$

In the subversive world of abstract categories, homomorphisms are called *functors*.

1.6   Groups

A *group* A is a semigroup (with identity) in which every element is invertible, so that

$$A = A^{-1} . \tag{1.6.1}$$

Groups are important in geometry and physics, in that they can be used to paint "algebraic pictures" of geometric objects and physical systems. The singleton $\mathbb{O}$ is a—rather trivial—group. Generally if A is a semigroup then the semigroup A^{-1} is a group: when we say that it is a subgroup of A we understand that the identity of A is included in, and therefore is the identity of, A^{-1}. Each of the number systems $\mathbb{Z}$, $\mathbb{Q}$, $\mathbb{R}$ and $\mathbb{C}$ are groups relative to *addition*, but none of them are groups relative to *multiplication*. However $\mathbb{Q}$, $\mathbb{R}$ and $\mathbb{C}$ are very nearly multiplicative groups:

$$A = \{0\} \cup A^{-1} . \tag{1.6.2}$$

Here we recall that additively the number 0 is the identity of each of the familiar number systems, and that paradoxically it behaves multiplicatively as a zero, while the number 1 is the multiplicative identity.

To qualify as a *subgroup*, a subset $K \subseteq A$ of a group A must be a subsemigroup, in the sense (1.5.7), for which also

$$x \in K \Longrightarrow x^{-1} \in K . \tag{1.6.3}$$

If $K \subseteq A$ is a subgroup then it generates an equivalence relation, in each of two ways: we can declare elements x, y to be "equivalent" provided

$$xy^{-1} \in K , \ alternatively \ if \ instead \ y^{-1}x \in K . \tag{1.6.4}$$

If the subgroup $K \subseteq A$ is *normal*, in the sense that

$$x \in A \Longrightarrow xK = Kx , \tag{1.6.5}$$

then these two relations coincide, and the *cosets* form another group A/K, with

$$(xK)(yK) = (xy)K \ (x, y \in K) . \tag{1.6.6}$$

This comes with a natural quotient homomorphism,

$$x \mapsto xK \ (A \to A/K) . \tag{1.6.7}$$

If A and B are groups then the direct sum semigroup $A \oplus B$ is also a group.

1.7 Rings

The number systems A are actually *rings*, in the sense that they support two semigroup operations, addition

$$(x, y) \mapsto x + y : A \times A \to A , \qquad (1.7.1)$$

which forms a commutative group, and multiplication

$$(x, y) \mapsto x \cdot y = xy , \qquad (1.7.2)$$

which is "distributed over" addition, in the sense that, if $x, y, z \in A$

$$x \cdot (y + z) = (x \cdot y) + (x \cdot z) \text{ and } (y + z) \cdot x = (y \cdot x) + (z \cdot x) . \qquad (1.7.3)$$

We shall write the additive inverse of $x \in A$ as $-x$, so that

$$x + (-x) = 0 = (-x) + x , \qquad (1.7.4)$$

and also write

$$x + (-y) = x - y . \qquad (1.7.5)$$

The reader is invited to deduce from the distributive law that the additive identity 0 is necessarily a multiplicative zero,

$$x \cdot 0 = 0 = 0 \cdot x , \qquad (1.7.6)$$

and that generally

$$x \cdot (-y) = -(x \cdot y) = (-x) \cdot y ; \ (-x) \cdot (-y) = x \cdot y . \qquad (1.7.7)$$

Generally when we discuss rings we treat them as their multiplicative semigroups, taking the additive structure for granted; thus in a ring the "inverse" will always be the multiplicative inverse.

A ring with the property (1.6.2) is called a *division ring*, and if it is also commutative is called a *field*. Thus the rationals, the reals and the complexes are all fields; the famous *quaternions* of Hamilton are a division ring.

If A and B are rings then so is the direct sum $A \oplus B$, where we define both products from the appropriate semigroup version (1.5.9). If A is a ring and X is a set then $A^X = \text{Map}(X, A)$ is also a ring, using both semigroup formulae (1.5.9).

If X is a commutative group (written additively) then the set $L(X) = L(X, X)$ of additive mappings from X to X is a ring, with

$$(S + T)(x) = (Sx) + (Tx) , \ (ST)(x) = S(T(x)) \ (x \in X ; \ S, T \in L(X, X)) : \qquad (1.7.8)$$

the reader is recommended to check the distributive laws. If J is a finite set and A is a ring then $A^{J \times J}$ is a ring, with

$$(S+T)_{ij} = S_{ij} + T_{ij} , \ (ST)_{ij} = \sum_{k \in J} T_{ik} S_{kj} \ (i, j \in J ; \ S, T \in A^{J \times J}) . \qquad (1.7.9)$$

If A is a ring and z is an "indeterminate" then the "polynomials"

$$A[z] = \{a_0 + a_1 z + \ldots + a_k z^k : k \in \mathbb{N}, \{a_j : j = 0, 1, \ldots, k\} \subseteq A\} \qquad (1.7.10)$$

form a ring; the reader is invited to write down sums and products.

If A is a ring then we recall that the set $\mathbf{2}^A$ of its subsets is a semigroup twice over: we can both add and multiply subsets as in (1.1.6). Note however that $\mathbf{2}^A$ is not in general another ring; we only get part of the distributive laws:

$$L \cdot (K + H) \subseteq (L \cdot K) + (L \cdot H) ; (K + H) \cdot L \subseteq (K \cdot L) + (H \cdot L) . \qquad (1.7.11)$$

In a ring or additive category, the *two-sided residial quotient* of H by K is given by

$$K : H = (L_K + R_K)^{-1}(H) \equiv \{x \in A : Kx + xK \subseteq H\} . \qquad (1.7.12)$$

If in particular

$$A \subseteq L(M) \equiv L_{\mathbb{Z}}(M) \qquad (1.7.13)$$

then M is said to be a left A module, and *simple* if there is implication

$$AN \subseteq N + N \subseteq N \subseteq M \Longrightarrow N \in \{O, M\} ; \qquad (1.7.14)$$

now it is *Schur's lemma* that

$$L_A(M) = \text{comm}(A) \ \textit{is a division ring} : \qquad (1.7.15)$$

for if $T \in \text{comm}(A) \subseteq L(M)$ then $N = T^{-1}(0)$ and $N = T(M)$ both satisfy (1.7.14), and hence either

$$T^{-1}(0) = O \ and \ T(M) = M$$

or

$$T^{-1}(0) = M \ and \ T(M) = O .$$

We also have the *Jacobson density theorem*: if $m \subseteq M^k$ is L_A *linearly independent*, in the sense that

$$b \in L_A(M)^k , \quad \sum_{j=1}^{k} b_j m_j = 0 \in M \Longrightarrow b = 0 , \tag{1.7.16}$$

then for arbitrary $m' \in M^k$ there is $a \in A$ for which

$$m' = am \in M^k . \tag{1.7.17}$$

1.8   Ideals

A *homomorphism* $T : A \to B$ of rings will be a semigroup homomorphism in both possible senses: relative both to addition and to multiplication. A subset $J \subseteq A$ of a semigroup is called a *left ideal*, respectively a *right ideal* if there is inclusion

$$AJ \subseteq J , \quad \textit{respectively } JA \subseteq J , \tag{1.8.1}$$

and a *two-sided ideal* if it is both a left and a right ideal. Obviously left and right ideals are therefore semigroups in their own right. When A is a ring we ask that also J is an additive subgroup, necessarily "normal"; now the quotient A/J is again a ring.

The ring A and the zero element $O = \{0\}$ are both left and right ideals, hence two-sided, and if $J \subseteq A$ is either a left or a right ideal then there is equivalence

$$A = J \Longleftrightarrow 1 \in J . \tag{1.8.2}$$

Ideals other than A are called *proper*, and it is an easy application, using (1.8.2), of Zorn's lemma that every proper left, (resp. right) ideal J is contained in a maximal (proper) left (resp. right) ideal M:

$$AJ \subseteq J \neq A \Longrightarrow J \subseteq M \in ML(A) . \tag{1.8.3}$$

The intersection $\bigcap ML(A)$ of all the maximal proper left ideals, and the intersection $\bigcap MR(A)$ of all the maximal proper right ideals, each coincide with what is called the (Jacobson)*Radical*:

$$\text{Rad}(A) = \{x \in A : 1 - Ax \subseteq A^{-1}\} = \{x \in A : 1 - xA \subseteq A^{-1}\} . \tag{1.8.4}$$

The fact that these two descriptions are the same is a consequence of what is sometimes called *Jacobson's lemma*: if $x, y \in A$ are arbitrary then there is equivalence

$$1 - xy \in A^{-1} \Longleftrightarrow 1 - yx \in A^{-1}. \tag{1.8.5}$$

In fact (1.8.5) holds separately for left and for right invertibility, and indeed the "relative regularity" of (1.2.13): of four separate implications we invite the reader to confirm that, if $x, y, w \in A$, then

$$w(1 - yx) = 1 \Longrightarrow (1 + xwy)(1 - xy) = 1, \tag{1.8.6}$$

which shows that $1 - yx \in A^{-1}_{left}$ implies $1 - xy \in A^{-1}_{left}$. We observe that the invertible group can be replaced by the left, and also by the right, invertibles, in the definition (1.8.4):

$$\mathrm{Rad}(A) = \{x \in A : 1 - Ax \subseteq A^{-1}_{left}\} = \{x \in A : 1 - xA \subseteq A^{-1}_{right}\}. \tag{1.8.7}$$

If for example $1 - Aa \subseteq A^{-1}_{left}$ then for arbitrary $a' \in A$ there is $a'' \in A$ for which

$$a''(1 - a'a) = 1, \implies a'' = 1 + (a''a')a \in A^{-1} \implies 1 - a'a \in A^{-1}. \tag{1.8.8}$$

If $J \subseteq A$ is a two-sided ideal then the quotient mapping $x \mapsto x + J$ is a homomorphism, from A to the quotient A/J; conversely if $T : A \to B$ is a homomorphism then the null space $T^{-1}(0) \subseteq A$ is a two-sided ideal. In particular the natural quotient

$$T : A \to B = A/\mathrm{Rad}(A) \tag{1.8.9}$$

has the Gelfand property (1.5.6). The ring A is described as *semi simple* when

$$\mathrm{Rad}(A) = O \equiv \{0\}; \tag{1.8.10}$$

in particular the quotient $A/\mathrm{Rad}(A)$ is always semi simple.

We remark that there are very few relatively regular radical elements:

$$A^{\cap} \cap \mathrm{Rad}(A) = \{0\}. \tag{1.8.11}$$

In particular if A is not semi simple then the quotient $a \mapsto a + \mathrm{Rad}(A)$ does not have "generalized permanence", in the sense of equality in (1.5.17).

If however $J \subseteq A$ is a two-sided ideal with the property

$$J \subseteq A^{\cap}, \tag{1.8.12}$$

then the quotient mapping $a \mapsto a + J$ has generalized permanence: to see this verify that

$$a - aca \in A^{\cap} \implies a \in A^{\cap} . \tag{1.8.13}$$

It is thus clear that spectral permanence is not in general sufficient for generalized permanence, and indeed spectral and generalized permanence together imply that a homomorphism is one one. It turns out that spectral permanence together with one one is also not sufficient for generalized permanence: indeed if

$$B_{left}^{-1} \neq B_{right}^{-1}$$

then there exist A and $T : A \to B$ for which $T : A \to B$ is one one with spectral but not generalized permanence. If indeed A is commutative, so that

$$A^{\cap} = SP(A) ,$$

then certainly, recalling (1.4.11),

$$T(A^{\cap}) \cap B_{left}^{-1} \neq \emptyset$$

guarantees the failure of generalized permanence. To achieve this we can take the natural embedding

$$T = J : \mathrm{comm}_B^2(a) \subseteq B$$

with

$$a \in B_{left}^{-1} \setminus B^{-1} .$$

Generally there is implication

    spectral permanence and simple permanence imply one one ,   (1.8.14)

while, conversely

    simple permanence and one one imply spectral permanence .   (1.8.15)

However, as we have just seen, spectral permanence and one one are not in general sufficient for simple permanence.

If $J \subseteq A$ is a two-sided ideal define its *inessential hull* by setting

$$\mathrm{Hull}(J) = \{a \in A : a + J \in \mathrm{Rad}(A/J)\} ; \tag{1.8.16}$$

thus

$$\text{Rad}(A) = \text{Hull}(O) . \tag{1.8.17}$$

The reader may like to verify the following extension of Jacobson's lemma (1.8.5): if

$$aba = aca \tag{1.8.18}$$

then

$$1 - ba \in A^{-1} \iff 1 - ac \in A^{-1} , \tag{1.8.19}$$

and similarly for left invertibles and for right invertibles.

The core of the argument (1.8.6) survives: indeed if (1.8.18) holds then we claim

$$c'(ac - 1) = 1 \implies a = c'(ac - 1)a = c'a(ba - 1)$$

$$\implies ba - 1 = bc'a(ba - 1) - 1$$

$$\implies (bc'a - 1)(ba - 1) = 1 .$$

This applies when $c = b$, and then continues to hold after interchanging a and b.

1.9   Fredholm Theory

When $T : A \to B$ is a homomorphism of rings then the semigroup of *Fredholm* elements (1.5.5) includes the sub semigroup

$$A^{-1} + T^{-1}(0) \subseteq T^{-1}B^{-1} \tag{1.9.1}$$

of *Weyl elements*, which in turn includes the *Browder elements*

$$A^{-1} +_{comm} T^{-1}(0) = \{c + d : c \in A^{-1}, d \in T^{-1}(0), cd = dc\} . \tag{1.9.2}$$

When $T : A \to B$ is onto then it is equivalent to the process of quotienting out the two-sided ideal $J = T^{-1}(0)$. If in particular $J \subseteq A^{\cap}$ is "completely regular" in the sense (1.8.12) then it is necessary and sufficient for $a \in A$ to be Fredholm that

$$a \in A^{\cap} ; \ L_a^{-1}(0) \subseteq J ; \ R_a^{-1}(0) \subseteq J . \tag{1.9.3}$$

If

$$J \subseteq K \subseteq A$$

then it is necessary and sufficient for the implication

$$a + K \in (A/K)^{-1} \implies a + J \in (A/J)^{-1} \tag{1.9.4}$$

that

$$K/J \subseteq \mathrm{Rad}(A/J) . \tag{1.9.5}$$

Equivalently, $K \subseteq \mathrm{Hull}(J)$, where $\mathrm{Hull}(J)$ is given by (1.8.16). In the category A of all linear operators between vector spaces, with the two-sided "ideal" J of *finite rank* operators, the conditions (1.9.3) say that $a = T : X \to Y$ is "spatially Fredholm" provided it has finite dimensional null space $T^{-1}(0) \subseteq X$ and also range $T(X) \subseteq Y$ of finite codimension:

$$\max(\dim T^{-1}(0), \dim(Y/T(X))) < \infty . \tag{1.9.6}$$

Now we define the *index* by setting

$$\mathrm{index}(T) = \dim T^{-1}(0) - \dim(Y/T(X)) . \tag{1.9.7}$$

For Fredholm operators $T : X \to Y$ and $S : Y \to Z$ there is now the "logarithmic law" of the index:

$$\mathrm{index}(ST) = \mathrm{index}(S) + \mathrm{index}(T) : \tag{1.9.8}$$

more generally for linear operators $T : X \to Y$ and $S : Y \to Z$ for which

$$T = TT'T ; \; S = SS'S ; \; ST = STUST$$

there is isomorphism

$$T^{-1}(0) \times S^{-1}(0) \times Z/ST(X) \cong (ST)^{-1}(0) \times Y/T(X) \times Z/S(Y) . \tag{1.9.9}$$

Operators $a : X \to Y$ which lie the left hand side of (1.9.1) are "spatially Weyl", those Fredholm operators for which the index is zero.

There are various kinds of "Riesz property", weaker than the Gelfand property, available to homomorphisms: we shall say that $T : A \to B$ is *completely regular* provided

$$T^{-1}(0) \subseteq A^{\cap} , \tag{1.9.10}$$

regular provided

$$A^{-1} + T^{-1}(0) \subseteq A^\cap , \tag{1.9.11}$$

and *decomposably regular* provided

$$A^{-1} + T^{-1}(0) \subseteq A^\cup . \tag{1.9.12}$$

We shall say that the homomorphism $T : A \to B$ has the *Riesz property* provided $A^{-1} + T^{-1}(0)$ is a subset of the polar elements of A; later we shall relax to the quasi polar elements of (4.3.12).

One way for "Fredholm" elements to have an "index" is that there be a "trace"

$$\text{tr} : T^{-1}(0) \to \mathbb{C} ,$$

for which, if $a \in A$ and $d, d' \in T^{-1}(0)$,

$$\text{tr}(d + d') = \text{tr}(d) + \text{tr}(d') ; \ \text{tr}(ad) = \text{tr}(da) ; \tag{1.9.13}$$

now the index of $a \in T^{-1}B^{-1}$ is to be given by :

$$\{1 - a'a, 1 - aa''\} \subseteq T^{-1}(0) \Longrightarrow \text{index}(a) = \text{tr}(aa' - a''a) . \tag{1.9.14}$$

The reader should check that this is well-defined, satisfies the logarithmic law (1.9.8), and is unchanged by $T^{-1}(0)$ perturbation. We shall describe $a \in A$ as "T-inessential" provided $a \in \text{Hull}(T^{-1}(0))$, as given by (1.8.18).

1.10 Exactness

We shall call the ordered pair $(b, a) \in A^2$ weakly exact if there is implication, for arbitrary $(u, v) \in A^2$,

$$va = 0 = bu \Longrightarrow vu = 0 , \tag{1.10.1}$$

whether or not it satisfies the *chain condition*

$$ba = 0, \tag{1.10.2}$$

and splitting exact if

$$1 \in Ab + aA . \tag{1.10.3}$$

Evidently there is implication

$$splitting\ exact \implies weakly\ exact\,, \tag{1.10.4}$$

while

$$weakly\ exact\ regular \implies splitting\ exact\,; \tag{1.10.5}$$

conversely

$$a\ splitting\ exact\ chain\ is\ regular\,. \tag{1.10.6}$$

Here $(b, a) \in A^2$ is called *regular* provided, in the notation of (1.2.13),

$$\{a, b\} \subseteq A^\cap\,. \tag{1.10.7}$$

For example

$$(0, a)\ splitting\ exact \iff a \in A^{-1}_{right} \tag{1.10.8}$$

and

$$(b, 0)\ splitting\ exact \iff b \in A^{-1}_{left}\,, \tag{1.10.9}$$

while

$$(0, a)\ weakly\ exact \iff a \in A^o_{right} \tag{1.10.10}$$

and

$$(b, 0)\ weakly\ exact \iff b \in A^o_{left}\,. \tag{1.10.11}$$

Generally regularity and products do not mix: neither of the implications

$$a \in A^\cap \implies a^2 \in A^\cap$$

and

$$a^2 \in A^\cap \implies a \in A^\cap$$

hold in general. However, when $(b, a) \in A^2$ is splitting exact we do have implication

$$ba \in A^\cap \iff \{a, b\} \subseteq A^\cap\,. \tag{1.10.12}$$

Indeed if $ba = bacba$ and $b'b + aa' = 1$ then

$$(1 - aa')a(1 - cba) = 0 = (1 - bac)b(1 - b'b) \,;$$

conversely if $b = bb^\wedge b$ and $a = aa^\wedge a$ and $b'b + aa' = 1$ then

$$(ba)^\wedge = a^\wedge b^\wedge \,.$$

When A is a ring then we shall say that $a \in A$ is *weakly, (resp. splitting), self exact* provided the pair (a, a) is weakly, (resp. splitting), exact, *weakly, (resp. splitting), n exact* if (a^n, a) is weakly, (resp. splitting), exact, and *weakly, (resp. splitting), hyper exact* provided it is weakly, (resp. splitting), n exact for arbitrary $n \in \mathbb{N}$.

Notice, for chains $(b, a) \in A^2$, how splitting exactness, weak exactness and regularity conform to the "democratic consensus" (1.2.17).

1.11 Block Structure

If G is a ring, with identity I, then an idempotent

$$Q = Q^2 \in G \tag{1.11.1}$$

induces *block structure*

$$G \cong \begin{pmatrix} A & M \\ N & B \end{pmatrix} = \{ T = \begin{pmatrix} a & m \\ n & m \end{pmatrix} : (a, m, n, b) \in A \times M \times N \times B \}, \tag{1.11.2}$$

where

$$A = QGQ \,, \quad M = QG(I - Q) \,, \quad N = (I - Q)GQ \,, \quad B = (I - Q)G(I - Q) \,. \tag{1.11.3}$$

Evidently A and B are themselves rings with identity, while M and N are bimodules over A and B; specifically M is a (left A, right B) bimodule, while N is a (left B, right A) bimodule, and there are also bilinear mappings

$$(m, n) \mapsto mn : M \times N \to A \,, \quad (m, n) \mapsto nm : M \times N \to B \,. \tag{1.11.4}$$

Multiplying generic elements $T \in G$ together like 2×2 matrices lays bare the structure. We can now write

$$I = \begin{pmatrix} 1 & 0 \\ 0 & 1 \end{pmatrix} \,; \quad Q = \begin{pmatrix} 1 & 0 \\ 0 & 0 \end{pmatrix} \,. \tag{1.11.5}$$

Necessary and sufficient for a generic element $T \in G$ to commute with the structural idempotent Q,

$$TQ = QT \,, \tag{1.11.6}$$

is that it be a *block diagonal*,

$$T = \begin{pmatrix} a & 0 \\ 0 & b \end{pmatrix} \in \begin{pmatrix} A & O \\ O & B \end{pmatrix} \subseteq G \,. \tag{1.11.7}$$

More generally there are upper and lower block triangles :

$$TQ = QTQ \Longleftrightarrow T = \begin{pmatrix} a & m \\ 0 & b \end{pmatrix} \in \begin{pmatrix} A & M \\ O & B \end{pmatrix} \tag{1.11.8}$$

and

$$QT = QTQ \Longleftrightarrow T = \begin{pmatrix} a & 0 \\ n & b \end{pmatrix} \in \begin{pmatrix} A & O \\ N & B \end{pmatrix} \,. \tag{1.11.9}$$

The invertibility of a block diagonal $T \in G$ is easily expressed in terms of $a \in A$ and $b \in B$:

$$T \in G^{-1} \Longleftrightarrow a \in A^{-1} \text{ and } b \in B^{-1} \,. \tag{1.11.10}$$

More generally if $T \in G$ is a block triangle, then the three statements

$$T \in G^{-1} \,; \; a \in A^{-1} \,; \; b \in B^{-1} \tag{1.11.11}$$

form another democratic consensus: each is a consequence of the conjunction of the other two. More generally still, this democratic consensus for (1.11.11) extends to "spectral triangles", T for which

$$1 - Mn \subseteq A^{-1} \text{ or } 1 - Nm \subseteq B^{-1} \,. \tag{1.11.12}$$

For example

$$(a \in A_{left}^{-1} \,, \; b \in B_{left}^{-1} \,, \; 1 - Mn \subseteq A_{left}^{-1}) \Longrightarrow T = \begin{pmatrix} a & m \\ n & b \end{pmatrix} \in G_{left}^{-1} \,, \tag{1.11.13}$$

and conversely

$$(T \in G_{left}^{-1} \,, \; 1 - Mn \subseteq A_{left}^{-1}) \Longrightarrow a \in A_{left}^{-1} \,, \tag{1.11.14}$$

while

$$(T \in G_{left}^{-1} , \ 1 - Mn \subseteq A_{left}^{-1} , \ a \in A_{right}^{-1}) \Longrightarrow b \in B_{left}^{-1} . \qquad (1.11.15)$$

To see this observe

$$a'a = 1 = a''(1 - a'mb'n) \in A , \ b'b = 1 \in B \Longrightarrow$$

$$\begin{pmatrix} a' & -a'mb' \\ 0 & b' \end{pmatrix} T = \begin{pmatrix} 1 - a'mb'n & 0 \\ b'n & 1 \end{pmatrix} \in G_{left}^{-1} ;$$

conversely

$$a'a + m'n = 1 \Longrightarrow a'a = 1 - m'n \in 1 + Mn \subseteq A_{left}^{-1} ,$$

while, using Jacobson's lemma (1.8.5),

$$n' = n'aa'' = -b'na'' ; \ \Longrightarrow b'b = 1 - n'm = 1 + b'na''m \in B_{left}^{-1}$$

$$\Longleftrightarrow 1 + na''mb' \in B_{left}^{-1} .$$

It follows therefore, that (1.11.10) continues to hold for "spectral diagonals": $T \in G$ for which

$$1 - Mn \subseteq A^{-1} \ and \ 1 - Nm \subseteq B^{-1} . \qquad (1.11.16)$$

If $T \in G$ is an upper spectral triangle then each of the following assertions implies its successor:

$$T \in G^{-1} \ and \ (a \in A_{right}^{-1} \ or \ b \in B_{left}^{-1}) ; \qquad (1.11.17)$$

$$a \in A^{-1} \ and \ b \in B^{-1} ; \qquad (1.11.18)$$

$$T \in G^{-1} ; \qquad (1.11.19)$$

$$a \in A_{left}^{-1} \ and \ b \in B_{right}^{-1} . \qquad (1.11.20)$$

If $T \in G$ then it is necessary and sufficient, for $T \in G^{-1}$, that there are $a' \in A$, $b' \in B$ and $n' \in N$ for which

$$a'a = 1 \in A ; \ bb' = 1 \in B \qquad (1.11.21)$$

with

$$1 - aa' = mn' \in A \text{ and } 1 - b'b = n'm \in B . \tag{1.11.22}$$

Condition (1.11.22) says that the idempotents $1 - aa' \in A$ and $1 - b'b \in B$ are *similar*. When $A = L(Y) \equiv L_{\mathbb{Z}}(Y)$ and $B = L(Z) \equiv L_{\mathbb{Z}}(Z)$ it says that their ranges are isomorphic. Necessary and sufficient, for $a \in A$, $b \in B$ and "M-radical" $n \in N$, that there exist $m \in M$ for which $T \in G^{-1}$ is that $a \in A_{left}^{-1}$ and $b \in B_{right}^{-1}$ with left and right inverses a' and b' for which the similarity condition (1.11.22) holds. When $A = L(Y)$ and $B = L(Z)$ this reduces to isomorphism

$$Y/a(Y) \cong b^{-1}(0) . \tag{1.11.23}$$

If an upper spectral triangle has a left inverse which is also an upper spectral triangle, then its diagonal elements will also have to be left invertible.

1.12 Linear Spaces and Algebras

A *linear space* over the field $\mathbb{K}$ is a commutative group, written additively, with an additional *scalar multiplication*

$$(\lambda, x) \mapsto \lambda x \ (\mathbb{K} \times X \to X) \tag{1.12.1}$$

for which

$$\begin{aligned} \lambda(x + y) = (\lambda x) + (\mu y) , \ (\lambda + \mu)x = (\lambda x) + (\mu x) , \\ (\lambda\mu)x = \lambda(\mu x) , \ 1x = x \ (x, y \in X , \ \lambda, \mu \in \mathbb{K}) \end{aligned} \tag{1.12.2}$$

The basic operation in a linear space is *linear combination*, where we write

$$\lambda x + \mu y = (\lambda x) + (\mu y) . \tag{1.12.3}$$

In nearly all cases the field $\mathbb{K}$ will be either the real field $\mathbb{R}$ or the complex field $\mathbb{C}$. A finite family $(x_j)_{j \in J}$ in a linear space X is described as *linearly independent* if there is implication, for arbitrary $(\lambda_j)_{j \in J}$ in the field $\mathbb{K}$,

$$\sum_{j \in J} \lambda_j x_j = 0 \in X \implies (\forall j \in J : \lambda_j = 0 \in \mathbb{K}) . \tag{1.12.4}$$

Algebras are both rings and linear spaces, with another commutative law:

$$(\lambda x)y = \lambda(xy) = x(\lambda y) \ (x, y \in A , \ \lambda, \mu \in \mathbb{K}) . \tag{1.12.5}$$

In effect, if $e \in A \neq \{0\}$ is the (multiplicative) identity then the subset $\mathbb{K}e \subseteq A$ is a copy of the number system $\mathbb{K}$. We will frequently write $e = 1$ for the identity in A, in deliberate confusion.

Direct sums $A \oplus B$, mapping spaces A^X, matrix rings $A^{J \times J}$ and polynomials $A[z]$ all become algebras when they are built out of algebras: the reader is invited to write down the appropriate scalar multiplication.

It is obvious that every complex algebra is also a real algebra; conversely there is a natural procedure for converting real algebras into complex, an obvious extension of the process of building the complex numbers from the reals. It turns out that we will almost always deal with complex algebras.

In a complex linear algebra there is a *functional calculus*, enabling us to define $f(a)$ for polynomials

$$f \in \mathrm{Poly} = \mathbb{C}[z] \tag{1.12.6}$$

and certain rational functions: if we write

$$z = I : \lambda \mapsto \lambda \; (\mathbb{C} \to \mathbb{C}) \tag{1.12.7}$$

for the fundamental complex coordinate we can define

$$(\alpha_0 + \alpha_1 z + \ldots + \alpha_k z^k)(a) = \alpha_0 + \alpha_1 a + \ldots + \alpha_k a^k , \tag{1.12.8}$$

and for polynomials p and q

$$q(a) \in A^{-1} \implies (p/q)(a) = p(a)q(a)^{-1} = q(a)^{-1}p(a) . \tag{1.12.9}$$

Evidently, for arbitrary polynomials $f, g \in \mathrm{Poly}$, and $a \in A$

$$(f + g)(a) = f(a) + g(a) ; \; (f \cdot g)(a) = f(a)g(a) ; \; (g \circ f)(a) = g(f(a)) . \tag{1.12.10}$$

Here of course

$$a^0 = 1 ; \; a^{k+1} = aa^k = a^k a .$$

This extends to rationals f and g, qualified by the necessary caveats: if $f = p/q$ and $g = r/s$ then we can write

$$f + g = (s \cdot p + q \cdot r)/(q \cdot s) , \; (f \cdot g) = (p \cdot r)/(q \cdot s) , \tag{1.12.11}$$

and recall that

$$\{q(a), s(a)\} \subseteq A^{-1} \implies q(a)s(a) \in A^{-1} , \tag{1.12.12}$$

which guarantees that if $f(a)$ and $g(a)$ both exist then so do $f(a) + g(a)$ and $f(a)g(a)$. On the other hand for $(g \circ f)(a)$ to exist it is necessary that $f(a)$ and then $g(f(a))$ both exist.

If $a \in A$ then the mapping

$$f \mapsto f(a) : \text{Poly} \mapsto A \qquad (1.12.13)$$

is a homomorphism, which extends to a larger algebra built from certain rational functions,

$$\{p/q : p, q \in \text{Poly}, q(a) \in A^{-1}\} . \qquad (1.12.14)$$

If A and B are linear algebras then a homomorphism $T : A \to B$ will be a ring homomorphism which also respects linear combination,

$$T(\lambda a + \mu b) = \lambda T(a) + \mu T(b) ,$$

and similarly for antihomomorphisms. An *involution* on a linear algebra will satisfy the conditions (1.5.21), with also

$$(\lambda a + \mu b)^* = \bar{\lambda} a^* + \bar{\mu} b^* . \qquad (1.12.15)$$

In a complex linear algebra a *hermitian* involution $a \mapsto a^*$ has the property

$$(a = a^* , \ \lambda \in \mathbb{C} \setminus \mathbb{R}) \Longrightarrow a - \lambda \in A^{-1} . \qquad (1.12.16)$$

We will occasionally meet another special antihomomorphism, the *adjugate*,

$$a \mapsto a^\sim = \text{adj}(a) : A \to A , \qquad (1.12.17)$$

for which

$$1^\sim = 1 ; \ (ba)^\sim = a^\sim b^\sim ; \ a^\sim a = aa^\sim = |a| \in \mathbb{K} \subseteq A , \qquad (1.12.18)$$

giving also the *determinant*

$$|a| = \det(a) \in \mathbb{K} . \qquad (1.12.19)$$

Evidently both the antihomomorphism *adj* and the homomorphism *det* have the Gelfand property (1.5.6).

Pairs of points give rise to *intervals*: : if $\{x, y\} \subseteq X$ we write

$$[x, y] = \{tx + (1 - t)y : 0 \le t \le 1\} , \qquad (1.12.20)$$

and then define the *convex hull* of $K \subseteq X$ by setting

$$\mathrm{cvx}(K) = \bigcap \{[x, y] : \{x, y\} \subseteq K\} ; \tag{1.12.21}$$

now we declare $K \subseteq X$ to be *convex* whenever

$$\mathrm{cvx}(K) \subseteq K . \tag{1.12.22}$$

Evidently

$$\{x, y\} \subseteq X \Longrightarrow [x, y] = \mathrm{cvx}(\{x, y\}) . \tag{1.12.23}$$

The *extreme points* of a convex set K are defined as the set

$$\mathrm{ext}(K) = \{x \in K : x \in [y, z] \subseteq K \Longrightarrow y = x = z\} . \tag{1.12.24}$$

For complex X the *absolutely convex hull* of $K \subseteq X$ is defined as the set

$$\mathrm{cvx}_{\mathbb{C}}(K) = \bigcup \{\lambda K + \mu K : |\lambda| + |\mu| \leq 1\} . \tag{1.12.25}$$

1.13   Linear Operators

If X and Y are linear spaces then a linear operator $T : X \to Y$ is a mapping which respects linear combination:

$$T(\lambda x + \mu y) = \lambda T(x) + \mu T(y) \tag{1.13.1}$$

We write

$$T \in L(X, Y) \equiv L_{\mathbb{K}}(X.Y) , \tag{1.13.2}$$

when (1.13.1) holds, and notice that $L(X, Y)$ is again a linear space.

If $T : X \to Y$ and $S : Y \to Z$ are linear operators then we say that the pair (S, T) is *exact* provided

$$S^{-1}(0) \subseteq T(X) , \tag{1.13.3}$$

again whether or not

$$ST = 0 : X \to Z . \tag{1.13.4}$$

When $X = Y = Z$ this agrees with the weak exactness (1.10.1) for the ring $A = L(X)$; indeed generally it is the extension of (1.10.1) to the *additive category* of linear mappings between vector spaces.

If X is a complex linear space then the set $A = L(X) \equiv L(X, X)$ of linear mappings from X to X is a linear algebra, in which everything is "relatively regular" in the sense (1.2.12), satisfying

$$A = A^\cap . \qquad (1.13.5)$$

This rests partly on the theory of *Hamel basis*: every basis for the null space $a^{-1}(0)$ and for the range $a(X)$ have extensions to X, and hence there are subspaces $Y \subseteq X$ and $Z \subseteq X$ for which

$$a^{-1}(0) + Y = X = a(X) + Z \; ; \; a^{-1}(0) \cap Y = \{0\} = a(X) \cap Z . \qquad (1.13.6)$$

These decompositions induce idempotent linear mappings $p, q \in A$ defined by setting, for each $x \in X$,

$$p(x) \in Y \text{ and } x - p(x) \in a^{-1}(0)$$

and

$$q(x) \in a(X) \text{ and } x - q(x) \in Z .$$

The reader should verify that $p^2 = p$ and $q^2 = q$ and

$$ap = a = qa \; ; \; a^{-1}(0) = p^{-1}(0) \; ; \; a(X) = q(X) .$$

Now if we define, for arbitrary $x, y \in X$,

$$b(a(x) + (1 - q)(y)) = p(x)$$

then $a = aba$. Since everything in A is regular it (1.3.11) follows

$$A_{left}^{-1} = A_{left}^o = \{a \in A : a^{-1}(0) = \{0\}\} , \qquad (1.13.7)$$

and

$$A_{right}^{-1} = A_{right}^o = \{a \in A : a(X) = X\} . \qquad (1.13.8)$$

More generally if A is a linear algebra then all this applies to the operators L_a and R_a of (1.3.1), which are in the algebra $L(A)$. In addition

$$a \in A_{left}^o \iff L_a \in L(A)_{left}^o \qquad (1.13.9)$$

and

$$a \in A^o_{right} \iff R_a \in L(A)^o_{left} . \tag{1.13.10}$$

Certainly if $a \in A^o_{left}$, then if $T \in L(A)$ there is implication

$$0 = L_a \circ T \in L(A) \iff \forall x : 0 = L_a(Tx) \implies \forall x : Tx = 0 \iff T = 0 .$$

This gives implication one way round in (1.13.9), and similarly (1.13.10). Conversely we need *linear functionals* $\varphi : X \to \mathbb{C}$, and *rank one* mappings

$$\varphi \odot y : x \mapsto \varphi(x)y \ (X \to X) . \tag{1.13.11}$$

Thus if $L_a \in L(A)^o_{left}$ and $x \in A$ and $\varphi \in L(A, \mathbb{C})$ then

$$ax = 0 \implies 0 = \varphi \odot ax = L_a \circ (\varphi \odot x) \implies \varphi \odot x = 0 ;$$

this for all φ implies $x = 0$. This is the other half of (1.13.9), and similarly the other half of (1.13.10).

An *invariant subspace* for $T \in L(X)$ is a subspace for which

$$T(Y) \subseteq Y \subseteq X , \tag{1.13.12}$$

giving rise to a *restriction operator* $T_Y : Y \to Y$, given by the formula

$$T_Y(y) = Ty \ (y \in Y) , \tag{1.13.13}$$

and a *quotient operator* $T_{/Y} : X/Y \to X/Y$, defined by setting

$$T_{/Y}(x + Y) = Tx + Y \ (x \in X) . \tag{1.13.14}$$

Analogous to left and right invertibility of products (1.2.4) and (1.2.5)

$$\left(T_Y, T_{/Y} \ one \ one \right) \implies T \ one \ one \implies T_Y \ one \ one \tag{1.13.15}$$

and

$$\left(T_Y, T_{/Y} \ onto \right) \implies T \ onto \implies T_{/Y} \ onto ; \tag{1.13.16}$$

conversely

$$\left(T \ one \ one , T_Y \ onto \right) \implies T_{/Y} \ one \ one \tag{1.13.17}$$

and

$$\left(T \ onto \ , \ T_{/Y} \ one \ one\right) \Longrightarrow T_Y \ onto \ . \tag{1.13.18}$$

An invariant subspace $T(Y) \subseteq Y \subseteq X$ for which

$$T^{-1}(0) \cap Y \subseteq T(Y) \tag{1.13.19}$$

will be called a *subplatform* for $T \in L(X)$; equivalently the pair $(T_Y, T_Y) \in L(Y)^2$ is exact.

Evidently the invertibility of

$$T \in L(X) \ , \ T_Y \in L(Y) \ , \ T_{/Y} \in L(X/Y)$$

conforms to the democratic consensus (1.2.17).

A representation, in the sense of a homomorphism $T : A \to L(X)$, is said to be *irreducible* provided there are no linear subspaces $Y \subseteq X$ other than $O = \{0\}$ and X for which $T(A)Y \subseteq Y$:

$$T(A)Y \subseteq Y \Longrightarrow Y \in \{O, X\} \ . \tag{1.13.20}$$

Equivalently, the linear space X is a simple left $T(A)$ module in the sense of (1.7.14); thus Schur's lemma (1.7.15) says that $\mathrm{comm}_B T(A) \subseteq B = L(X)$ is a division ring, and we also again have the Jacobson density theorem (1.7.17).

If we specialize the discussion of §1.4 to the algebra $A = L(X)$ then, with again $p = p^2$ and $q = q^2$ satisfying (1.4.15) $(1 - q)(1 - p) = 0$ as in (1.4.15) then

$$q(X) = p^{-1}(0) + (pq)(X) \tag{1.13.21}$$

and

$$p^{-1}(0) \cap (pq)(X) = O \ ; \tag{1.13.22}$$

We do not, however, claim

$$q = (1 - p) + pq \ . \tag{1.13.23}$$

1.14 Hyper Exactness

When $A = L(X)$ is the linear operators on a vector space then weak and splitting exactness coincide, and hyperexactness reduces to inclusion

$$T^{-1}(0) \subseteq T^{\infty}(X) \ ; \tag{1.14.1}$$

equivalently

$$T^{-\infty}(0) \subseteq T(X).\tag{1.14.2}$$

Here

$$T^{\infty}(X) = \bigcap_{n=1}^{\infty} T^n(X)\tag{1.14.3}$$

and

$$T^{-\infty}(0) = \bigcup_{n=1}^{\infty} T^{-n}(0)\tag{1.14.4}$$

are respectively the *hyperrange* and the *hyperkernel* of $T \in L(X)$. We shall say that $a \in A$ is *Kato invertible* in A provided it is relatively regular and (splitting) hyper exact. Equivalence (1.14.1)$\Longleftrightarrow$(1.14.2) follows from the rather general implications

$$(V, TU), (T, U)\ splitting\ exact \implies (VT, U)\ splitting\ exact\tag{1.14.5}$$

and

$$(VT, U), (V, T)\ splitting\ exact \implies (V, TU)\ splitting\ exact.\tag{1.14.6}$$

Indeed

$$V'V + (TU)U' = I = T'T + UU'' \implies I = T'(V'V + TUU')T + UU'''.$$

When

$$T^{-n}(0) = T^{-\infty}(0)\tag{1.14.7}$$

then $T \in L(X)$ is said to have *ascent* $\leq n$, while if

$$T^n(X) = T^{\infty}(X)\tag{1.14.8}$$

then T has *descent* $\leq n$. Necessary and sufficient for $T \in L(X)$ to be *simply polar* in the sense (1.4.6) is that it have both ascent and descent ≤ 1.

The hyperrange (1.14.3) is the tip of an iceberg: we can define $T^{\alpha}(X)$ for arbitrary *ordinals* α if we set

$$T^{\alpha+1}(X) = T\ T^{\alpha}(X)\tag{1.14.9}$$

and define, for limit ordinals α,

$$T^{\alpha}(X) = \bigcap_{\beta < \alpha} T^{\beta}(X) . \tag{1.14.10}$$

Similarly

$$T^{-(\alpha+1)}(0) = T^{-1} T^{-\alpha}(0) \tag{1.14.11}$$

and, for limit ordinals,

$$T^{-\alpha}(0) = \bigcup_{\beta < \alpha} T^{-\beta}(0) . \tag{1.14.12}$$

We shall declare the pair $(S, T) \in L(X)^2$ middle Taylor non singular if there is inclusion

$$\left(-S \quad T\right)^{-1}(0) \subseteq \binom{T}{S} X : \tag{1.14.13}$$

evidently we are declaring a certain other pair of linear operators, derived from the pair (S, T), to be exact in the sense (1.13.3). Necessary and sufficient for middle non singularity (1.14.13) are the following three conditions:

$$S^{-1}(0) \subseteq T \, S^{-1}(0) ; \tag{1.14.14}$$

$$T^{-1}(0) \subseteq S \, T^{-1}(0) ; \tag{1.14.15}$$

$$S(X) \cap T(X) \subseteq (ST)(TS - ST)^{-1}(0) . \tag{1.14.16}$$

It is, further, a consequence of (1.14.14)–(1.14.16), that

$$(ST)^{-1}(0) + (TS)^{-1}(0) \subseteq S^{-1}(0) + T^{-1}(0) . \tag{1.14.17}$$

It now follows that, in the presence of middle non singularity (1.14.13), there is equality

$$\begin{aligned} (S^{-1}0 \cap SX) + (T^{-1}0 \cap TX) = \\ ((ST)^{-1}0 + (TS)^{-1}0) \cap (ST)X \cap (TS)X \end{aligned} \tag{1.14.18}$$

and

$$\begin{aligned} (S^{-1}0 + SX) \cap (T^{-1}0 + TX) = \\ ((ST)^{-1}0 + (TS)^{-1}0) + ((ST)X \cap (TS)X) \end{aligned} . \tag{1.14.19}$$

This in turn implies

$$\max(\text{ascent}(S), \text{ascent}(T)) \leq 1 \iff \\ \max(\text{ascent}(ST), \text{ascent}(TS)) \leq 1 \qquad (1.14.20)$$

and

$$\max(\text{descent}(S), \text{descent}(T)) \leq 1 \iff \\ \max(\text{descent}(ST), \text{descent}(TS)) \leq 1 . \qquad (1.14.21)$$

This of course looks a lot simpler when S and T commute.

Topology

2

The language of spectral theory may be algebra, but its environment is *topological algebra*. Topology on the one hand underlies geometry; for us it is more relevant that it also underlies the theory of limits and analysis.

2.1 Topological Spaces

A *topological space* is a set which carries an additional structure, its "topology". There are various ways of specifying a topology on a set: the most economical, if not the most intuitive, is by specifying the *open sets*. For a set $G \subseteq 2^X$ of subsets of a set X to be the open sets of a topology requires three conditions:

$$\{\emptyset, X\} \subseteq G \, ; \tag{2.1.1}$$

$$\{U, V\} \subseteq G \Longrightarrow U \cap V \in G \, ; \tag{2.1.2}$$

$$H \subseteq G \Longrightarrow \bigcup_{U \in H} U \in G \, . \tag{2.1.3}$$

The associated theory of limits will be described in terms of *neighbourhoods*: if $x \in X$ and $U \subseteq X$ we shall say that

$$U \in \mathrm{Nbd}(x) \Longleftrightarrow \exists V \in G \, : \, x \in V \subseteq U \, . \tag{2.1.4}$$

Conversely the set G of open sets is determined by the "neighbourhood mapping": we define the *interior* of a subset $K \subseteq X$ to be

$$\mathrm{int}\, K = \{x \in X : K \in \mathrm{Nbd}(x)\} \, , \tag{2.1.5}$$

© The Author(s), under exclusive license to Springer Nature Switzerland AG 2023
R. Harte, *Spectral Mapping Theorems*,
https://doi.org/10.1007/978-3-031-13917-8_2

and can now verify (exercise!) that

$$K \in G \iff K \subseteq \text{int } K \ . \tag{2.1.6}$$

We add two further definitions: we declare that the *closure* of a set $K \subseteq X$ is

$$\text{cl } K = \{x \in X : \forall U \in \text{Nbd}(x) \ : \ K \cap U \neq \emptyset\} \ , \tag{2.1.7}$$

and that a set $K \subseteq X$ is *closed* if and only if

$$\text{cl } K \subseteq K \ . \tag{2.1.8}$$

It can now be proved (exercise! exercise!) that, for arbitrary $K \subseteq X$,

$$X \setminus \text{cl } K = \text{int}(X \setminus K) \ , \tag{2.1.9}$$

and that

$$K \ closed \iff X \setminus K \ open \ . \tag{2.1.10}$$

This of course depends on the properties (2.1.1)–(2.1.3) of the set of open sets. It is perfectly possible to make (2.1.9) the *definition* of the "closure" and then prove (2.1.7) as a consequence.

If X and Y are topological spaces then there is a natural *product topology* on the cartesian product $X \times Y$, for which

$$W \in \text{Nbd}(x, y) \iff \exists U \in \text{Nbd}(x), V \in \text{Nbd}(y) \ : \ U \times V \subseteq W \ . \tag{2.1.11}$$

2.2   Continuity

If $f : X \to Y$ is a mapping between topological spaces and if $x \in X$ then we shall say that f is *continuous* at x if there is implication

$$\forall V \in \text{Nbd}(f(x)) \ , \ \exists U \in \text{Nbd}(x) \ : \ f(U) \subseteq V \ . \tag{2.2.1}$$

Equivalently

$$\forall V \in \text{Nbd}(f(x)) \ , \ \exists U \in \text{Nbd}(x) \ : \ U \subseteq f^{-1}(V) \ . \tag{2.2.2}$$

We note two "opposite" properties at this point: we shall say that f is *open* at x if instead there is implication

$$\forall U \in \text{Nbd}(x) \ , \ \exists V \in \text{Nbd}(f(x)) \ : \ V \subseteq f(U) \ , \tag{2.2.3}$$

and that f is *bounded below* at x if instead

$$\forall U \in \mathrm{Nbd}(x) \,, \; \exists V \in \mathrm{Nbd}(f(x)) \,:\, f^{-1}(V) \subseteq U \,. \qquad (2.2.4)$$

The reader may like to verify that if $f : X \to Y$ and $g : Y \to Z$ are mappings between topological spaces, and $x \in X$, then there is implication

$$f \text{ continuous at } x \,, \; g \text{ continuous at } f(x) \implies g \circ f \text{ continuous at } x \,, \qquad (2.2.5)$$

$$f \text{ open at } x \,, \; g \text{ open at } f(x) \implies g \circ f \text{ open at } x \qquad (2.2.6)$$

and

$$f \text{ bounded below at } x \,, \; g \text{ bounded below at } f(x) \implies g \circ f \text{ bounded below at } x \,; \qquad (2.2.7)$$

conversely there is implication

$$g \circ f \text{ open at } x \implies g \text{ open at } f(x) \,; \qquad (2.2.8)$$

and

$$g \circ f \text{ bounded below at } x \implies f \text{ bounded below at } x. \qquad (2.2.9)$$

2.3   Limits

If $x = (x_n)$ is a *sequence* in X, formally a mapping from $\mathbb{N}$ to X, and if $y \in X$, then we shall say that x *converges to* y provided there is a family N_U in $\mathbb{N}$, indexed by the neighbourhoods $U \in \mathrm{Nbd}(y)$, for which there is implication, for arbitrary $n \in \mathbb{N}$ and $U \in \mathrm{Nbd}(y)$,

$$n \geq N_U \implies x_n \in U \,. \qquad (2.3.1)$$

Convergence is not restricted to sequences: we can replace the natural numbers $\mathbb{N}$ by a "directed set" and speak of "generalized sequences". We can also speak of *limits*: if X and Y are topological spaces, $x \in X$ and $y \in Y$, and if $f : K \to Y$ is defined on a subset $K \subseteq X$, then we shall say that

$$f(w) \to y \text{ as } w \,\rightarrow\, x \; (w \in K) \qquad (2.3.2)$$

provided that

$$\forall V \in \mathrm{Nbd}(y) \,, \; \exists U \in \mathrm{Nbd}(x) \,:\, x \neq w \in U_{\cap}K \implies f(w) \in V \,. \qquad (2.3.3)$$

We note that the value $f(x)$ is irrelevant to (2.3.2), and indeed "$f(x)$" need not even be defined. For such limits to be meaningful however it is necessary that the point $x \in \text{acc}(K) \subseteq X$ be an *accumulation point* of the set K, in the sense that

$$U \in \text{Nbd}(x) \implies (U \cap K) \setminus \{x\} \neq \emptyset . \qquad (2.3.4)$$

It is also helpful if the topology of Y is *separated*, in the sense that for each $y \in Y$

$$\bigcap \{V : V \in \text{Nbd}(y)\} = \{y\} . \qquad (2.3.5)$$

For example if $X = \mathbb{N} \cup \{\infty\}$ and we define, for each $n \in \mathbb{N}$,

$$U \in \text{Nbd}(n) \iff n \in U$$

and

$$U \in \text{Nbd}(\infty) \iff \exists m \in \mathbb{N} : \{\infty\} \cup (\mathbb{N} + m) \subseteq U$$

then ∞ is an accumulation point of $\mathbb{N}$, and (2.3.3) reduces to (2.3.1) with $f(n) = x_n$. Alternatively, with $K = X$ and $y = f(x)$, the condition (2.3.3) is equivalent to the condition (2.2.1) for continuity at x.

2.4   Metric Spaces

One way for a set to become a topological space is that it play host to a *distance function*

$$\text{dist} : X \times X \to \mathbb{R} ,$$

satisfying the conditions, for arbitrary x, y, z in X,

$$0 \leq \text{dist}(x, y) ; \qquad (2.4.1)$$

$$\text{dist}(y, x) = \text{dist}(x, y) ; \qquad (2.4.2)$$

$$\text{dist}(x, z) \leq \text{dist}(x, y) + \text{dist}(y, z) ; \qquad (2.4.3)$$

$$\text{dist}(x, y) = 0 \implies y = x . \qquad (2.4.4)$$

When a distance function is prescribed on a set X then it becomes a *metric space*. A distance function gives rise to neighbourhoods: if $x \in X$ define

$$\text{Ball}_\delta(x) = \{y \in X : \text{dist}(x, y) \leq \delta\} , \qquad (2.4.5)$$

and if also $U \subseteq X$ declare

$$U \in \text{Nbd}(x) \iff \exists \delta > 0 : \text{Ball}_\delta(x) \subseteq U . \tag{2.4.6}$$

The reader can easily verify the open set conditions (2.1.1)–(2.1.3), using (2.1.6) to identify the family of open sets: the crucial property of distance functions which makes this work is the *triangle inequality* (2.4.3). The symmetry condition (2.4.2) is a natural convenience to assume; the implication (2.4.4) guarantees that the topology is separated in the sense (2.3.5).

Convergence and continuity mean the same as for general topological spaces, but in metric spaces there is the new idea of *Cauchy sequences* and *completeness*. A sequence (x_n) in a metric space X is said to be *Cauchy* if it has the property

$$\text{dist}(x_n, x_m) \to 0 \, ((n, m) \to (\infty, \infty)) : \tag{2.4.7}$$

formally there is a mapping $\varepsilon \mapsto N_\varepsilon$ from the strictly positive reals to the natural numbers for which there is implication, for arbitrary n, m in $\mathbb{N}$,

$$\forall \varepsilon > 0 : N_\varepsilon \leq \min(m, n) \implies \text{dist}(x_n, x_m) < \varepsilon . \tag{2.4.8}$$

It is an easy but fundamental exercise to verify that every sequence (x_n) which converges in the sense (2.3.1) is Cauchy; if conversely all Cauchy sequences are convergent the metric space X is said to be *complete*.

Evidently, if $K \subseteq X$ for a complete space X,

$$K \text{ complete} \iff K \text{ closed} . \tag{2.4.9}$$

The reader is invited to formulate the analogue of the Cauchy condition for the mappings of (2.3.2), and to verify that if they converge in the sense (2.3.3) then they are also Cauchy. It turns out that if a metric space Y is complete then every such "Cauchy" mapping also has a limit. In complete metric spaces one can know that certain sequences are convergent without any detailed information about their limits. At a certain level of elementary calculus this might be said for example about the real sequence (x_n) given by setting

$$x_n = \sum_{r=1}^{n} r^{-2} \, (n \in \mathbb{N}) . \tag{2.4.10}$$

For another example define $x_n \in \{0, 1\}$ to be 0 provided the equation

$$a^k = b^k + c^k \tag{2.4.11}$$

has no solutions in $\{1, 2, \ldots n\}$ unless $k \leq 2$, and $x_n = 1$ else.

If $K \subseteq X$ then int cl(K) is "where K is dense". The *Baire category theorem* says that a complete metric space X cannot be the union of countably many nowhere dense subsets $K_n \subseteq X$:

$$\bigcup_{n=1}^{\infty} \text{int cl}(K_n) = \emptyset \Longrightarrow \text{int} \bigcup_{n=1}^{\infty} K_n = \emptyset \ . \qquad (2.4.12)$$

The *contraction mapping principle* guarantees that, on a complete metric space X, certain well-behaved mappings have *fixed points*: if there is $0 \leq k < 1$ for which, for arbitrary $x, y \in X$,

$$\text{dist}(f(x), f(y)) \leq k \, \text{dist}(x, y) \ , \qquad (2.4.13)$$

then there is a unique point $e \in X$ for which

$$f(e) = e \ . \qquad (2.4.14)$$

The condition (2.4.13) ensures that, for arbitrary $x \in X$, the sequence $(f^n(x))$ is Cauchy, hence by completeness convergent; now $e = \lim_n f^n(x)$ obviously satisfies (2.4.14), and if also $f(e') = e'$ then

$$0 \leq \text{dist}(e', e) = \text{dist}(f(e'), f(e)) \leq k \, \text{dist}(e'.e) \ . \qquad (2.4.15)$$

Metric spaces are also "normal", in the sense that whenever

$$\text{cl } K_0 \subseteq \text{int } K_1 \subseteq X \ , \qquad (2.4.16)$$

there is also $K_{1/2} \subseteq X$ for which

$$\text{cl } K_0 \subseteq \text{int } K_{1/2} \subseteq \text{cl } K_{1/2} \subseteq \text{int } K_1 \ . \qquad (2.4.17)$$

It follows that $K_0 \subseteq K_1$ satisfying (2.4.16) can be embedded in a *relief map* $(K_\lambda)_{\lambda \in \Lambda}$ of X, indexed by a dense subset $\Lambda \subseteq [0, 1]$ for which, whenever $0 \leq \lambda < \mu \leq 1$, $K_\lambda \subseteq K_\mu$ satisfies (2.4.16). The *Urysohn function* of the relief map K_λ is the mapping $K^\vee : X \to [0, 1]$ given by setting

$$K^\vee(x) = \inf\{\lambda \in \Lambda : x \in K_\lambda\} \ , \qquad (2.4.18)$$

with in addition

$$K^\vee(x) = 1 \ (x \notin \bigcup_\lambda K_\lambda) \ ; \ = 0 \ (x \in \bigcap_\lambda K_\lambda) \ . \qquad (2.4.19)$$

Urysohn's Lemma says that the Urysohn function of a relief is continuous. In turn, Urysohn's lemma gives rise to "partitions of unity"; every open cover (U_j) of a compact metric space $X = \bigcup_j U_j$ supports a *subordinate* partition of unity (p_j) in $C(X, [0, 1])$:

$$\forall j : X \setminus U_j \subseteq p_j^{-1}(0) \; ; \; \sum_j p_j \equiv 1 \; . \tag{2.4.20}$$

For a specific formula set

$$x \in X \Longrightarrow p_j(x) = \frac{q_j(x)}{\sum_k q_k(x)} \; , \tag{2.4.21}$$

where

$$q_j(x) = \text{dist}(x, X \setminus U_j) \; . \tag{2.4.22}$$

The *Cantor intersection theorem* says that it is necessary and sufficient, for X to be complete, that whenever $(K_j)_{j \in J}$ is a nest of nonempty closed subsets of X, there is implication

$$\inf_{j \in J} \text{diam}(K_j) = 0 \Longrightarrow \bigcap_{j \in J} K_j \neq \emptyset \; ; \tag{2.4.23}$$

necessarily, of course, that intersection will be a singleton.

Here, if $x \in X$ and $K \subseteq X$,

$$\text{dist}(x, K) = \inf_{y \in K} \text{dist}(x, y) \tag{2.4.24}$$

and

$$\text{diam}(K) = \sup_{x, y \in K} \text{dist}(x, y) \; . \tag{2.4.25}$$

2.5 Compactness

A subset $K \subseteq X$ of a topological space is said to be *compact* if, whenever

$$U_x \in \text{Nbd}(x) \; (x \in K) \; , \tag{2.5.1}$$

there is finite $K_0 \subseteq K$ for which

$$K \subseteq \bigcup_{x \in K_0} U_x \; . \tag{2.5.2}$$

Equivalently, every open cover for K has a finite subcover: if $H \subseteq G$ is a set of open subsets of X for which

$$K \cap \bigcap_{U \in H} U = \emptyset \qquad (2.5.3)$$

then there must be a finite subset $H_0 \subseteq H$ for which

$$K \cap \bigcap_{U \in H_0} U = \emptyset . \qquad (2.5.4)$$

If $(X_\lambda)_{\lambda \in \Lambda}$ is a family of compact spaces X_λ then it is *Tychenoff's theorem* that the cartesian product

$$Y = \prod_{\lambda \in \Lambda} X_\lambda \qquad (2.5.5)$$

is compact: here, in the topology of Y, $U \in \mathrm{Nbd}(y)$ means that

$$\forall \lambda \in \Lambda : U_\lambda \in \mathrm{Nbd}(y_\lambda) \qquad (2.5.6)$$

while

$$\#\{\lambda \in \Lambda : U_\lambda \neq X_\lambda\} < \infty : \qquad (2.5.7)$$

in words only finitely many of the projections U_λ differ from the whole space X_λ.

The elements $y \in Y$ of (2.5.5) are *choice functions*: if none of the factors X_λ are empty it needs *Zorn's lemma* to be sure that the product Y is also not empty.

When X is a metric space we shall say that $K \subseteq X$ is *totally bounded* if, for arbitrary $\delta > 0$, there is a *finite δ-net*, $K_\delta \subseteq K$ for which

$$K \subseteq \bigcup_{x \in K_\delta} \mathrm{Ball}_\delta(x) . \qquad (2.5.8)$$

When $K \subseteq X$ for a metric space X there is equivalence

$$K \text{ compact} \iff K \text{ totally bounded and complete} . \qquad (2.5.9)$$

When in (2.5.5) we have

$$\lambda \in \Lambda \implies X_\lambda = E , \qquad (2.5.10)$$

then the cartesian product $Y = E^\Lambda$ is the set of all mappings from Λ to E, with the topology of *pointwise convergence* .

More generally, X is *locally compact* provided that whenever $x \in X$ there is implication

$$U \in \mathrm{Nbd}(x) \implies \exists \; compact \; V \in \mathrm{Nbd}(x) \, ,$$

for which $V \subseteq U$.

Like metric spaces, locally compact Hausdorff spaces are also "normal", in the sense of (2.4.14) and (2.4.15); thus also Urysohn's lemma applies.

2.6   Boundaries, Hulls and Accumulation Points

A subset $K \subseteq X$ of a topological space is said to be *connected* if it has no *disconnections*, in the sense of pairs of open sets $U \subseteq X$, $V \subseteq X$ for which

$$K \subseteq U \cup V \, ; \; K \cap U \cap V = \emptyset \, ; \; K \cap U \neq \emptyset \neq K \cap V \, . \tag{2.6.1}$$

For example the connected subsets $J \subseteq \mathbb{R}$ are the *intervals*:

$$\{a, b\} \subseteq J \implies [a, b] \subseteq J \tag{2.6.2}$$

whenever $a \leq b$ in J. If K and H are subsets of X then

$$K \; connected \, , \; K \subseteq H \subseteq \mathrm{cl}(K) \implies H \; connected \tag{2.6.3}$$

and

$$K \, , \; H \; connected \, , \; K \cap H \neq \emptyset \implies K \cup H \; connected \, . \tag{2.6.4}$$

If $K \subseteq X$ and $H \subseteq Y$ then

$$K \; connected \; in \; X \, , \; H \; connected \; in \; Y \implies K \times H \; connected \; in \; X \times Y \, . \tag{2.6.5}$$

If $\varphi : X \to Y$ is a continuous mapping then

$$\varphi \; continuous \; on \; K \, , \; K \; connected \; in \; X \implies \varphi(K) \; connected \; in \; Y \, . \tag{2.6.6}$$

For example if $X = \mathbb{R}$ then (2.6.6), with (2.6.2), is known as the *intermediate value theorem*. When $x \in K \subseteq X$ then the *connected component* of x in K,

$$\mathrm{Comp}_x K = \bigcap \{U : x \in U \subseteq K; \, U \; connected\} \subseteq K \tag{2.6.7}$$

is evidently connected in X. The connected components form a *partition* of the set K: if $\{s, t\} \subseteq K$ there is implication

$$\mathrm{Comp}_s K \neq \mathrm{Comp}_t K \Longrightarrow \mathrm{Comp}_s K \cap \mathrm{Comp}_t K = \emptyset . \tag{2.6.8}$$

If $K \subseteq X$ and $t \in X \setminus K$ then the complement in X of $\mathrm{Comp}_t K$ will be known as the *connected hull* of K with respect to t:

$$X \setminus \eta_t K = \mathrm{Comp}_t (X \setminus K) . \tag{2.6.9}$$

If a *hole* in K with respect to $t \in X$ is a component in $X \setminus K$ other than $X \setminus \eta_t K$ then we obtain $\eta_t K$ by "filling in the holes". Evidently

$$\eta_t \eta_t K = \eta_t K , \tag{2.6.10}$$

and

$$H \subseteq K \Longrightarrow \eta_t H \subseteq \eta_t K . \tag{2.6.11}$$

If $K \subseteq X$ is closed and if X is *locally connected*, in the sense that

$$U \in \mathrm{Nbd}(t) \Longrightarrow t \in V \subseteq U \tag{2.6.12}$$

for some connected $V \in \mathrm{Nbd}(t)$ then

$$\eta_t K = \mathrm{cl}\, \eta_t K . \tag{2.6.13}$$

For compact $K \subseteq X$, when $X = \mathbb{C}$ is the complex plane, we shall write

$$\eta K = \eta_\infty K = \eta_t K \text{ with } |t| > \sup |K| . \tag{2.6.14}$$

There is interaction between the connected hull and the topological boundary: if H, K, L are subsets of X then

$$H \subseteq K , \ \partial K \subseteq H \cup L \Longrightarrow \partial K \subseteq (\partial H) \cup L , \tag{2.6.15}$$

and if H and K are closed then

$$\left(\partial K \subseteq H \text{ and } \partial H \subseteq K\right) \Longrightarrow \partial(K \cup H) \subseteq (\partial K) \cap (\partial H) \subseteq \partial(K \cap H) \subseteq K \cap H . \tag{2.6.16}$$

If X is locally connected and H and K are closed subsets of $X \setminus \{t\}$ then

$$\partial K \subseteq H \subseteq K \Longrightarrow H \subseteq K \subseteq \eta_t H \Longleftrightarrow \partial \eta_t K \subseteq H \subseteq K \tag{2.6.17}$$

and

$$\partial \eta_t K \subseteq \partial K \subseteq K \subseteq \eta_t K = \eta_t \partial K . \qquad (2.6.18)$$

Indeed if $H \subseteq K$ and $\partial K \subseteq H \cup L$ then

$$\partial K \subseteq (H \cup L) \cap (X \setminus \text{int } H) \subseteq (\partial H) \cup L .$$

Also if $L \subseteq X$ is connected then

$$(\partial K) \cap L = \emptyset \Longrightarrow \left(K \subseteq L \text{ or } L \subseteq K \text{ or } K \cap L \neq \emptyset \right) :$$

otherwise (int K, $X \setminus \text{cl } K$) would be a disconnection of L in X.

If H and K are subsets of X then, recalling the *accumulation points* acc(K) of (2.3.4), the *isolated points* of K are given by

$$\text{iso}(K) = K \setminus \text{acc}(K) . \qquad (2.6.19)$$

$$H \subseteq K \Longrightarrow \text{acc } H \subseteq \text{acc } K .$$

If X is locally connected then

$$\text{iso}(K) \subseteq \partial K$$

and hence also

$$\text{acc } \partial K = \partial \text{acc}(K) \ \textit{and} \ \text{iso } \partial K = \text{iso}(K) . \qquad (2.6.20)$$

Hence if X is locally connected and $H \subseteq K \subseteq X \setminus \{t\}$ then

$$\eta_t K = \eta_t \text{acc } K \cup \textit{iso } K \qquad (2.6.21)$$

and

$$\partial K \subseteq H \cup \text{iso } K \Longrightarrow K \subseteq \eta_t H \cup \text{iso } K \Longrightarrow \partial \eta_t K \subseteq H \cup \text{iso } K . \qquad (2.6.22)$$

In general, when $X = \mathbb{C}$, there is inclusion

$$\partial \eta K \subseteq \partial K \subseteq \eta K = \eta \partial K . \qquad (2.6.23)$$

If $\lambda \in \partial K \setminus \partial \eta K$ then there are sequences (λ_n) in $\mathbb{C}$ and (H_n) of connected components of $X \setminus K$ for which

$$\lambda_n \to \lambda \ (n \to \infty) \ \textit{and} \ \lambda_n \in \partial H_n \ (n \in \mathbb{N}) . \qquad (2.6.24)$$

If $f = p/q$ is non constant rational with $q^{-1}(0) \cap K = \emptyset$ then

$$\partial f(K) \subseteq f(\partial K) \qquad (2.6.25)$$

and

$$f(\eta K) \subseteq \eta f(K) . \qquad (2.6.26)$$

If $K \subseteq X$ is compact and $f : K \to Y$ is continuous then

$$\mathrm{acc}\, f(K) \subseteq f\, \mathrm{acc}\, K \subseteq \big(\mathrm{acc}\, f(K)\big) \cup \big(f(K) \setminus f(K'_f)\big) , \qquad (2.6.27)$$

where

$$K'_f = \{\lambda \in K : \exists\, U \in \mathrm{Nbd}(\lambda) : U \cap f^{-1}(\lambda) = \{\lambda\}\} \qquad (2.6.28)$$

consists of the points of K at which f is *locally one one*. There is also equality

$$\mathrm{acc}(K \times H) = \big(\mathrm{acc}\, K \times H\big) \cup \big(K \times \mathrm{acc}\, H\big) . \qquad (2.6.29)$$

Dually

$$\mathrm{iso}\, f(K) \subseteq f(\mathrm{iso}\, K) \cup \big(f(K) \setminus f(K'_f)\big) \qquad (2.6.30)$$

and

$$\mathrm{iso}(K \times H) = \mathrm{iso}\, K \times \mathrm{iso}\, H . \qquad (2.6.31)$$

For compact subsets of the complex plane $\mathbb{C}$ we observe

$$0 \notin \mathrm{acc}(K) \cup \mathrm{acc}(H) \Longrightarrow 0 \notin \mathrm{acc}(KH) : \qquad (2.6.32)$$

for if

$$0 \neq \lambda \in K \Longrightarrow |\lambda| > \varepsilon$$

and

$$0 \neq \mu \in K \Longrightarrow |\mu| > \delta$$

then

$$0 \neq \lambda\mu \in KH \Longrightarrow |\lambda\mu| > \varepsilon\delta .$$

2.7   Connectedness and Homotopy

If X and Y are topological spaces then continuous functions $f, g \in C(X, Y)$ are said to be *homotopic* if there is a continuous mapping

$$H \in C([0, 1] \times X, Y) \qquad (2.7.1)$$

for which

$$H(0, \cdot) = f(\cdot), \; H(1, \cdot) = g(\cdot) : \qquad (2.7.2)$$

when this happens we shall sometimes write $f \sim g$, and write

$$(h_t)_{0 \leq t \leq 1} = H(t, \cdot)_{0 \leq t \leq 1}$$

for the *homotopy* between $h_0 = f$ and $h_1 = g$. If $f \in C(X, Y)$ is homotopic to a constant we shall say that it is *contractible*. A topological space X is called contractible when the identity $I : X \to X$ is homotopic to a constant: for example normed spaces, and convex subsets of normed spaces are contractible, while bounded linear operators are contractible mappings. If $f : X \to Y$ and $g : Y \to Z$ are continuous then

$$(f \; contractible \; or \; g \; contractible) \implies g \circ f \; contractible. \qquad (2.7.3)$$

It follows that every continuous mapping in or out of a contractible space is contractible. The space X is said to be *arc-wise connected* if every pair of mappings into X from the one-point space $\{0\}$ are homotopic: here the homotopy (h_t) is referred to as an *arc*. Evidently

$$X \; contractible \implies X \, arc - wise \; connected \implies X \, connected. \qquad (2.7.4)$$

In partial converse, if X is "locally arc-wise connected" then

$$K \subseteq X \; open \; and \; connected \implies K \, arc - wise \; connected. \qquad (2.7.5)$$

The *circle*, boundary of the closed unit disc $\mathbb{D}$,

$$\mathbb{S} = \partial \mathbb{D} = e^{i\mathbb{R}} \subseteq \mathbb{C} \qquad (2.7.6)$$

is an example of a space which is not contractible. If we call a space X *simply connected* when every pair of continuous mappings from the circle $\mathbb{S}$ to X are homotopic then the circle itself also fails to be simply connected. Indeed if X is a topological space then it is necessary and sufficient for a continuous mapping

$\varphi : \mathbb{S} \to X$ to be contractible that it have a continuous extension

$$\varphi^{\wedge} : \mathbb{D} \to X . \tag{2.7.7}$$

In the opposite direction, it is necessary and sufficient for a continuous mapping $\varphi : X \to \mathbb{S}$ to be contractible that it have a continuous "lift"

$$\varphi^{\vee} : X \to \mathbb{R} . \tag{2.7.8}$$

where

$$\varphi = \mathrm{ex}_\pi \circ \varphi^{\vee} . \tag{2.7.9}$$

Here we define

$$\mathrm{ex}_\pi : \mathbb{R} \to \mathbb{S} , \; \lg_\pi : \mathbb{S} \setminus \{-1\} \to \mathbb{R}$$

by setting

$$\mathrm{ex}_\pi (t) = e^{2\pi i t} \; (t \in \mathbb{R}) ; \; \lg_\pi (e^{2\pi i t}) = t \; (-1 < 2t < 1) .$$

In particular each of three conditions are separately equivalent to contractibility for $\varphi \in C(\mathbb{S}, \mathbb{S})$: the extension, the lift, and the vanishing of the *winding number*, or "degree",

$$\mathrm{index}(\varphi) = \varphi_*(1) - \varphi_*(0) , \tag{2.7.10}$$

where φ_* is obtained as a lift:

$$\varphi_* = (\varphi \circ \mathrm{ex}_\pi)^{\vee} \tag{2.7.11}$$

For example

$$\mathrm{index}(z) = 1 , \tag{2.7.12}$$

where $z : \mathbb{C} \to \mathbb{C}$ is the complex co-ordinate,

$$\mathrm{index}(\psi \cdot \varphi) = \mathrm{index}(\psi) + \mathrm{index}(\varphi) \tag{2.7.13}$$

and

$$\mathrm{index}(\psi \circ \varphi) = \mathrm{index}(\psi) \cdot \mathrm{index}(\varphi) . \tag{2.7.14}$$

The *Brouwer fixed point theorem* says that if $f : \mathbb{D} \to \mathbb{D}$ is continuous then it has a fixed point:

$$f \in C(\mathbb{D}, \mathbb{D}) \implies \exists e \in \mathbb{D} : f(e) = e . \qquad (2.7.15)$$

More generally if also $\varphi : \mathbb{D} \to \mathbb{D}$ is continuous and satisfies

$$\varphi(\mathbb{S}) \subseteq \mathbb{S} \qquad (2.7.16)$$

then

$$\mathrm{index}(\varphi_{\mathbb{S}}) \neq 0 \implies \exists \lambda \in \mathbb{D} : f(\lambda) = \varphi(\lambda) . \qquad (2.7.17)$$

Here $\varphi_{\mathbb{S}}$ is the restriction of φ to $\mathbb{S}$; if (2.7.15) fails then we can extend $\varphi_{\mathbb{S}}$ to continuous $\psi : \mathbb{D} \to \mathbb{S}$ by taking

$$\psi(\lambda) = f(\lambda) + t(\varphi(\lambda) - f(\lambda)) \in \mathbb{S}$$

to be the point where the line from $f(\lambda)$ to $\varphi(\lambda)$ meets the circle $\mathbb{S}$.

We can also see that if $\varphi \in C(\mathbb{S}, \mathbb{S})$ is contractible then it cannot have the *antipodal property*

$$\varphi(-z) \equiv -\varphi(z) ; \qquad (2.7.18)$$

to the contrary there must (*Borsuk-Ulam*) be $\lambda \in \mathbb{S}$ for which

$$\varphi(-\lambda) = \varphi(\lambda) . \qquad (2.7.19)$$

Spin-off includes the *fundamental theorem of algebra* :

$$p = a_0 + a_1 z + \ldots + z^n \implies \exists \lambda \in \mathbb{C} : p(\lambda) = 0 . \qquad (2.7.20)$$

To see this take $\varphi = z^n$ in (2.7.17), and then set

$$q(z) \equiv k^{-n} p(kz) \text{ with } |a_0| + |a_1|k + \ldots + |a_{n-1}|k^{n-1} < k^n ; \qquad (2.7.21)$$

we have

$$f \equiv z^n - q \implies f(\mathbb{D}) \subseteq \mathbb{D} , \qquad (2.7.22)$$

and then by (2.7.17) there is $\mu \in \mathbb{D}$ for which

$$f(\mu) = \mu^n ; \iff p(k\mu) = 0 . \qquad (2.7.23)$$

2.8   Disconnectedness

The most profoundly disconnected topological spaces are the *discrete* spaces, X for which

$$x \in X \Longrightarrow \{x\} \in \text{Nbd}(x) \; ; \tag{2.8.1}$$

equivalently

$$X = \text{iso}(X) \; . \tag{2.8.2}$$

More generally X is totally disconnected iff

$$x \in X \Longrightarrow \{x\} = \text{Comp}(x) \; ; \tag{2.8.3}$$

equivalently

$$X = \text{iso}^{\sim}(X) \; . \tag{2.8.4}$$

For example both the rationals $\mathbb{Q}$ and the irrationals $\mathbb{R} \setminus \mathbb{Q}$ are totally disconnected. Generally the quotient space $[X]_{\sim}$ generated by the equivalence relation for which

$$x' \sim x \Longleftrightarrow x' \in \text{Comp}(x) \tag{2.8.5}$$

is totally disconnected. The space X is said to be *extremally disonnected* if there is implication

$$K = \text{int}(K) \subseteq X \Longrightarrow \text{cl}(K) = \text{int cl}(K) \; . \tag{2.8.6}$$

For example an extremally disconnected Hausdorff space is totally disconnected; in its *cofinite topology* an infinite set is both connected and extremally disconnected.

The *Cantor ternary set*

$$C_{\infty} = \bigcap_{n=0}^{\infty} C_n \tag{2.8.7}$$

is totally disconnected, where, for each $n \in \infty$,

$$C_0 = [0, 1] \; ; \; 3C_{n+1} = C_n \cup (2 + C_n) \; . \tag{2.8.8}$$

We can access C_{∞} with "self-similar transformations"

$$T_L : x \mapsto x/3 \; ; \; T_R : x \mapsto (2 + x)/3 \; ;$$

now

$$C_{n+1} = T_L(C_n) \cup T_R(C_n) .$$

We have

$$[0, 1] \setminus C_\infty = \bigcup_{n=1}^{\infty} \bigcup_{k=0}^{3^n-1}](3k + 1)/3, (3k + 2)/3[, \qquad (2.8.9)$$

systematically removing middle thirds:

$$[(3k)/(3n + 1), (3k + 3)/(3n + 1)] \setminus](3k + 1)/(3n + 1), (3k + 2)/(3n + 1)[.$$

C_∞ is uncountable: $f(C_\infty) = [0, 1]$ is onto, where

$$\alpha \in \{0, 2\}^{\mathbb{N}} \implies f(\sum_{k=1}^{\infty} \alpha_k 3^{-k}) = \sum_{k=1}^{\infty} (1/2)\alpha_k 2^{-k} . \qquad (2.8.10)$$

Returning to the self-similar transformations,

$$T_L(C_\infty) \cong T_R(C_\infty) \cong C_\infty = T_L(C_\infty) \cup T_R(C_\infty) .$$

We have homeomorphism

$$C_\infty \cong \{0, 1\}^{\mathbb{N}} ;$$

now C_∞ is *homogeneous*, in the sense that if $\{x, y\} \subseteq C_\infty$ is arbitrary there is $h \in C(C_\infty, C_\infty)$ for which $h(x) = y$. Indeed define $h' : \{0, 1\}^{\mathbb{N}} \to \{0, 1\}^{\mathbb{N}}$ by setting

$$h'([u]_2) = [u + x + y]_2 .$$

C_∞ is uncountable, but has Lebesgue measure zero, and is also *perfect*, in the sense that

$$C_\infty = \text{acc } C_\infty ; \text{ int } C_\infty = \emptyset .$$

It is also true that $C_\infty \cong \mathbb{Z}_p$ is homeomorphic to the "p-adic integers", with

$$k = \min\{n \in \mathbb{N} : x_n \neq y_n\} \implies \text{dist}(x, y) = 2^{-k} .$$

Topological Algebra

3

When the same set carries both algebraic and topological structure then it is good if they are "compatible": this usually means that the algebraic operations are continuous.

3.1 Topological Semigroups

A *topological semigroup* is a semigroup A with a topology for which the mapping

$$* : (x, y) \mapsto xy : A \times A \to A \text{ is continuous} \tag{3.1.1}$$

at each point $(x, y) \in A \times A$. A *topological ring* is both additively and multiplicatively a topological semigroup, and a *topological algebra* has the additional property that scalar multiplication

$$(\lambda, x) \mapsto \lambda x \ (\mathbb{K} \times A \to A)$$

is everywhere continuous. Both addition and multiplication are continuous in this sense for each of the familiar number systems, in their usual topologies. For the number systems, and in general for well-behaved topological semigroups A, the invertible group is open,

$$A^{-1} = \text{int } A^{-1} \subseteq A , \tag{3.1.2}$$

and the inversion process continuous: for each point $a \in A^{-1}$,

$$x \mapsto x^{-1} \text{ is continuous at } a . \tag{3.1.3}$$

It is even part of the definition of "topological group" that the inversion map is continuous.

R. Harte, *Spectral Mapping Theorems*,
https://doi.org/10.1007/978-3-031-13917-8_3

The topology of a topological ring is determined by the set Nbd(0) of neighbour-hoods of the origin, since by addition continuity

$$\text{Nbd}(x) = x + \text{Nbd}(0) \ (x \in A) \,. \tag{3.1.4}$$

Even without a distance function the idea of "Cauchy" sequences and generalized sequences persists in topological rings: the reader is invited to interpret the condition (2.4.7) in this context.

A *topological vector space* is not only a topological group, with continuous addition, but also (jointly) continuous scalar multiplication: the mapping

$$(\lambda, x) \mapsto \lambda x : \mathbb{K} \times X \to X \tag{3.1.5}$$

is continuous. A *locally convex* space has the additional property that every neighbourhood of the origin includes a convex neighbourhood of the origin:

$$\forall U \in \text{Nbd}(0) \ \exists V \in \text{Nbd}(0) : V = \text{cvx}(V) \subseteq U \,. \tag{3.1.6}$$

3.2 Spectral Topology

If A is a ring then the algebra of the ring itself also generates a topology on A which we shall refer to as the *spectral topology*, where $Nbd(x) = x + Nbd(0)$ and $U \in Nbd(0)$ iff there is finite $J \subseteq A$ for which there is implication

$$1 - Jx \subseteq A^{-1} \Longrightarrow x \in U \,. \tag{3.2.1}$$

The reader is invited to describe the "spectral closure"

$$Cl(K) \subseteq A \tag{3.2.2}$$

of a subset $K \subseteq A$, what it means for a sequence (x_n) is A to converge to an element $y \in A$ in this topology, and indeed to carefully check that the spectral topology really is a topology: the "spectrally open sets" satisfy (2.1.1)–(2.1.3). Check also that in this topology we now indeed have a topological ring, in which inversion is continuous. Indeed it is necessary and sufficient, for a topological ring A to have an open invertible group A^{-1}, that the mapping

$$x \mapsto x : A \to A$$

is continuous, from the given topology to the spectral topology. In terms of closure this says that there is inclusion

$$cl(K) \subseteq Cl(K) \subseteq A \,, \tag{3.2.3}$$

for arbitrary $K \subseteq A$; while in terms of neighbourhoods

$$Nbd(x) \subseteq \text{Nbd}(x) \, , \tag{3.2.4}$$

for arbitrary $x \in A$.

It follows, in particular, that the spectral topology of a ring makes it a topological ring, with indeed an open invertible group A^{-1} with continuous inversion $x \mapsto x^{-1} : A^{-1} \rightarrow A$:

$$Cl(A \setminus A^{-1}) \subseteq A \setminus A^{-1} \tag{3.2.5}$$

and, for arbitrary $K \subseteq A^{-1}$,

$$(A^{-1} \cap Cl(K))^{-1} \subseteq Cl(K^{-1}) \, . \tag{3.2.6}$$

We remark that there is equality

$$A_{left}^{-1} \cap Cl(A_{right}^{-1}) = A^{-1} = A_{right}^{-1} \cap Cl(A_{left}^{-1}) \, . \tag{3.2.7}$$

There is also inclusion

$$A^{\cap} \cap Cl(A^{-1}) \subseteq A^{\cup} \, , \tag{3.2.8}$$

with equality if and only if there is implication

$$a^2 = a \Longrightarrow a \in Cl(A^{-1}) \, . \tag{3.2.9}$$

3.3   Normed Algebra

When $X = A$ is a commutative group, written additively, then one way to get a distance function is by means of a *norm*

$$\| \cdot \| : A \rightarrow \mathbb{R} \, ,$$

satisfying, for each x, y in A,

$$0 \leq \|x\| \, ; \tag{3.3.1}$$

$$\| - x \| - \|x\| \, ; \tag{3.3.2}$$

$$\|x + y\| \leq \|x\| + \|y\| \, ; \tag{3.3.3}$$

$$\|x\| = 0 \Longrightarrow x = 0 \, . \tag{3.3.4}$$

The induced distance function is given by

$$\text{dist}(x, y) = \|y - x\| \ (x, y \in X) ; \tag{3.3.5}$$

conversely $\|x\| = \text{dist}(0, x)$. Norms are usually only introduced on linear spaces, where they satisfy the additional condition

$$\|\lambda x\| \le |\lambda| \, \|x\| \ (\lambda \in \mathbb{K}, x \in A) ; \tag{3.3.6}$$

on rings, in particular algebras, it is assumed

$$\|xy\| \le \|x\| \, \|y\| \ (x, y \in A) \tag{3.3.7}$$

with also $\|1\| = 1$. The familiar *absolute value* $|\lambda|$ of a real or complex number λ is an example.

A normed algebra element $a \in A$ will be described as *quasinilpotent* whenever

$$\|a^n\|^{1/n} \to 0 \ (n \to \infty) . \tag{3.3.8}$$

Normed algebras and normed linear spaces are of course locally convex, in the sense (3.1.6).

3.4   Banach Algebra

A normed algebra which is complete is known as a *Banach algebra*. Banach algebras have the crucial property that the invertible group is open, and inversion is continuous: they satisfy the conditions (3.1.2) and (3.1.3). This flows from a seemingly inoffensive observation: if $x \in A$ there is implication

$$\|x\| < 1 \implies 1 - x \in A^{-1} \ with \ \|(1 - x)^{-1}\| \le (1 - \|x\|)^{-1} . \tag{3.4.1}$$

To see this consider the sequence (y_n) in A defined by setting

$$y_n = \sum_{r=0}^{n} x^r = 1 + x + \ldots + x^n \ (n \in \mathbb{N}) . \tag{3.4.2}$$

This turns out to be a Cauchy sequence: if $k \in \mathbb{N}$

$$\|y_{n+k} - y_n\| \to 0 \ (n \to \infty) . \tag{3.4.3}$$

This is the numerical argument about the convergence of the geometric series: if n, k are in $\mathbb{N}$ then

$$\|y_{n+k} - y_n\| = \|x^{n+1} + x^{n+2} + \ldots + x^{n+k}\| \leq \|x^n\| \, \|x + \ldots + x^k\|$$

$$\leq \|x^n\| \frac{1 - \|x\|^{k+1}}{1 - \|x\|} \leq \frac{\|x\|^n}{1 - \|x\|} \to 0 \, (n \to \infty) \,.$$

At the same time another sequence visibly converges:

$$y_n(1 - x) = (1 - x)y_n = 1 - x^{n+1} \to 1 \, (n \to \infty) \,.$$

There is therefore implication, for arbitrary $w \in A$,

$$y_n \to w \, (n \to \infty) \implies w(1 - x) = 1 = (1 - x)w \,. \tag{3.4.4}$$

It follows that the invertible group is an open set: if $x \in A^{-1}$ and $y \in A$ there is implication

$$\|x^{-1}\| \, \|y - x\| < 1 \implies x - y = x(1 - x^{-1}y) \in A^{-1}A^{-1} \subseteq A^{-1} \,,$$

remembering the product inequality (3.3.7). Continuity of the inverse (3.1.3) is now easy: if $x, y \in A^{-1}$ then

$$y^{-1} - x^{-1} = y^{-1}(x - y)x^{-1} = (y^{-1} - x^{-1})(x - y) + x^{-1}(y - x)x^{-1}$$

and hence

$$(1 - \|x - y\| \, \|x^{-1}\|)\|y^{-1} - x^{-1}\| \leq \|x^{-1}\| \, \|x - y\| \, \|x^{-1}\| \,;$$

thus

$$2\|y - x\| \, \|x^{-1}\| < 1 \implies \|y^{-1} - x^{-1}\| < 2\|x^{-1}\|^2\|y - x\| \to 0 \, (y \to x) \,.$$

We notice one more consequence of the inoffensive (3.4.1): if $K \subseteq A$ is arbitrary, $J \subseteq A$ is finite and if $x \in \text{cl}(K)$ is in its norm closure, then there is implication

$$\exists \, x' \in K \,:\, 1 - J(x' - x) \subseteq A^{-1} \,. \tag{3.4.5}$$

In words, the norm closure of a subset $K \subseteq A$ is a subset of its spectral closure. Equivalently, the identity mapping $I : A \to A$ is continuous if the left hand A has the norm topology while the right hand A has the spectral topology of (3.2.1).

In very much the same vein we have the *square root lemma*: if $k \in \mathbb{R}$, in particular with $k = 1/2$, there is with a binomial expansion, implication

$$\|x\| < 1 \Longrightarrow \exists y \in A \; : \; y^k = 1 - x \; . \tag{3.4.6}$$

Alternatively, there is a recurrence relation (*Newton's method*)

$$y_0 = 1; \, 2y_{n+1} = y_n + y_n^{-1}(1 - x) \; .$$

The really important Banach algebras A carry *involutions* $a \mapsto a^*$, which, in addition to the linear algebra conditions (1.5.21) and (1.12.15), are also bounded, and satisfy the "B* condition"

$$\|a^*a\| = \|a\|^2 \; . \tag{3.4.7}$$

A Banach algebra with such an involution is known as a "C* algebra": morally, we believe it would have been appropriate to call it a *Hilbert algebra*. The fundamental examples are $A = C(\Omega)$, the continuous functions on a compact Hausdorff space Ω, with

$$a^*(t) = \overline{a(t)} \; (t \in X) \; ,$$

and the algebra $A = B(X)$ on a *Hilbert space* X. A Hilbert space is a Banach space whose norm

$$\|x\| = \langle x; x \rangle^{1/2} \tag{3.4.8}$$

is given by an *inner product*

$$(x, y) \mapsto \langle x; y \rangle \; (X \times X \to \mathbb{C}) \tag{3.4.9}$$

which by definition satisfies

$$0 \leq \langle x; x \rangle \; ; \tag{3.4.10}$$

$$\langle \lambda x + \lambda' x'; y \rangle = \lambda \langle x; y \rangle + \lambda' \langle x'; y \rangle \; ; \tag{3.4.11}$$

$$\langle y; x \rangle = \langle x; y \rangle^- \; ; \tag{3.4.12}$$

together with implication

$$\langle x; x \rangle = 0 \in \mathbb{C} \Longrightarrow x = 0 \in X \; . \tag{3.4.13}$$

The involution $a \mapsto a^*$ on such $B(X)$ is given by the formula

$$\langle a^*x; y \rangle = \langle x; ay \rangle . \tag{3.4.14}$$

In particular the B* condition (3.4.7) gives cancellation (1.5.23), and also the hermitian property (1.5.28). In a C* algebra A it turns out that whenever an element has a generalized inverse then it also has a Moore-Penrose inverse: in the notation of (1.5.26) there is equality

$$A^\cap = A^\dagger . \tag{3.4.15}$$

It also holds that an isometric C* homomorphism $T : A \to B$ of C* algebras always has both spectral permanence (1.5.6), and generalized permanence, in the sense of equality in (1.5.17).

Necessary and sufficient that the norm on a Banach space X is given by an inner product in the sense (3.4.9) is that it have the following *quadratic property*: for arbitrary $x, y \in X$,

$$\|x + y\|^2 + \|x - y\|^2 \leq 2\|x\|^2 + 2\|y\|^2 ; \tag{3.4.16}$$

if (3.4.16) holds we can, by "*polarization*" define, in both real and complex spaces,

$$4 \operatorname{Re}\langle x; y \rangle = \|x + y\|^2 - \|x - y\|^2 . \tag{3.4.17}$$

For real spaces we now take $\langle x; y \rangle = \operatorname{Re}\langle x; y \rangle$; for complex spaces (*cf* (3.6.3) below)

$$\langle x; y \rangle = \operatorname{Re}\langle x; y \rangle - i\operatorname{Re}\langle ix; y \rangle . \tag{3.4.18}$$

The fundamental metric property of Hilbert spaces X is the following *nearest point theorem*: if

$$\emptyset \neq K = \operatorname{cl} \operatorname{cvx}(K) \subseteq X , \tag{3.4.19}$$

then for arbitrary $x \in X$ there is a uniquely determined point $E_K(x) \in X$ for which

$$E_K(x) \in K \ and \ \|x - E_K(x)\| = \operatorname{dist}(x, K) . \tag{3.4.20}$$

The mapping $E_K : X \to X$ is idempotent and continuous; when in particular $K \subseteq X$ is also a linear subspace, then $E_K \in B(X)$ is also linear, hence bounded, and obtained by "dropping the perpendicular" from x to K. It follows that a closed subspace $K \subseteq X$ is always *complemented*:

$$K + H = X ; \ K \cap H = O \equiv \{0\} , \tag{3.4.21}$$

for another (closed) subspace, the "orthogonal complement" $H = K^\perp \subseteq X$:

$$K = E_K(X) \; ; \; H = (I - E_K)(X) = E_K^{-1}(0) = K^\perp \; . \tag{3.4.22}$$

The B* condition (3.4.8) extends to the whole category of bounded Hilbert space operators: if X and Y are Hilbert spaces and $T \in BL(X, Y)$ we define $T^* \in BL(Y, X)$ by setting

$$\langle Tx; y \rangle_Y = \langle x; T^* y \rangle_X \; . \tag{3.4.23}$$

When $Y = X$ then we have equivalence

$$T^* = T \iff \{\langle Tx; x \rangle : x \in X\} \subseteq \mathbb{R} \; , \tag{3.4.24}$$

and we shall declare $T \in B(X)$ to be *positive* provided

$$\{\langle Tx; x \rangle : x \in X\} \subseteq [0, \infty) \subseteq \mathbb{R} \; . \tag{3.4.25}$$

Generally if $T \in BL(X, Y)$ then $T^*T \in B(X)$ is now "positive", and if also $S \in B(X)$ is positive then so is $T^*T + S$ and, for arbitrary $x \in X$,

$$\|Tx\|^2 \le \|x\| \; \|(T^*T + S)x\| \; . \tag{3.4.26}$$

In particular, there is, recalling (1.5.29), implication

$$(S \text{ positive and } T \text{ left invertible}) \implies T^*T + S \text{ invertible} \; . \tag{3.4.27}$$

This complemented property of closed subspaces has spin-off: whenever $T \in B(X)$ has closed range, then it also has a (bounded) generalized inverse:

$$T(X) = \text{cl } T(X) \implies T \in B(X)^\cap \; . \tag{3.4.28}$$

It therefore also follows, between Hilbert spaces, that

$$T \text{ bounded below} \implies T \text{ left invertible} \; , \tag{3.4.29}$$

and

$$T \text{ onto} \implies T \text{ right invertible} \; . \tag{3.4.30}$$

This complemented subspace property persists in Banach spaces X which are "Hilbert equivalent", in the sense that there are $m > 0$ and $M > 0$ and another, quadratic, norm $\| \cdot \|'$ for which, for arbitrary $x \in X$,

$$m\|x\| \le \|x\|' \le M\|x\| \; . \tag{3.4.31}$$

Conversely it is *Lindenstrauss' theorem* that whenever a Banach space fails to satisfy (3.4.31), then it has uncomplemented closed subspaces. The simplest concrete example would be

$$c_0 \subseteq \ell_\infty . \tag{3.4.32}$$

This is demonstrated by showing that for arbitrary (f_n) in ℓ_∞^* there is implication

$$c_0 \subseteq \bigcap_{n=1}^{\infty} f_n^{-1}(0) \Longrightarrow \bigcap_{n=1}^{\infty} f_n^{-1}(0) \nsubseteq c_0 ; \tag{3.4.33}$$

now if there were $P = P^2 \in B(\ell_\infty)$ with $c_0 = P(\ell_\infty)$, we would have, with $\pi_n(x) = x_n \ (n \in \mathbb{N})$,

$$f_n = \pi_n(I - P) \Longrightarrow c_0 = \bigcap_{n=1}^{\infty} f_n^{-1}(0) . \tag{3.4.34}$$

We shall see (5.1.26) below that the defining "B* condition" (3.4.7) of C* algebras implies something apparently stronger: if $a, b \in A$ then

$$\|a\|^2 \le \|a^*a + b^*b\| .$$

The cancellation property (1.5.23), and the hermitian property (1.5.28), hold; also, unless $A = O \equiv \{0\}$,

$$a^*a - \|a\|^2 \notin A^{-1} . \tag{3.4.35}$$

We remark that

$$a = a^* \notin A^{-1} \Longrightarrow \inf_{\|u\| \ge 1} \|au\| = \inf_{\|u\| \ge 1} \|ua\| = 0 , \tag{3.4.36}$$

while

$$\inf_{\|u\| \ge 1} \|au\| > 0 \Longrightarrow a \in A_{left}^{-1} \Longrightarrow a^*a \in A^{-1} . \tag{3.4.37}$$

3.5 Bounded Operators

If X and Y are normed linear spaces then a linear mapping $T : X \to Y$ will be continuous at the point $x \in X$ if and only if it is continuous at $0 \in X$: this is clear from (3.1.4). It turns out that necessary and sufficient that it be continuous at $0 \in X$ is that it be *bounded*, in the sense that there exists $k > 0$ in $\mathbb{R}$ with the property that

$$\forall x \in X : \|Tx\| \le k\|x\| . \tag{3.5.1}$$

It also turns out that the infimum of all $k > 0$ satisfying (3.5.1) is also a supremum:

$$\|T\| = \sup\{\|Tx\| : \|x\| \leq 1\} . \tag{3.5.2}$$

This—as we have taken the liberty of building into the notation—is indeed a norm, defined on the linear subspace

$$BL(X, Y) \subseteq L(X, Y)$$

of continuous linear operators from X to Y. The same arguments will show that $T \in L(X, Y)$ will be "bounded below" at $x \in X$, in the sense (2.2.4), if and only if it is bounded below at 0, and that necessary and sufficient will be that there is $k > 0$ for which

$$\forall x \in X : \|Tx\| \geq k\|x\| . \tag{3.5.3}$$

This time the supremum of all $k > 0$ satisfying (3.5.3) is also an infimum:

$$m(T) = \inf\{\|Tx\| : \|x\| \geq 1\} . \tag{3.5.4}$$

Similarly T will be "open" at $x \in X$, in the sense (2.2.3), if and only if it is open at 0, for which it is necessary and sufficient that there is $k > 0$ for which

$$\forall y \in Y : y \in \{Tx : \|x\| \leq k\|y\|\} . \tag{3.5.5}$$

We shall describe T satisfying (3.5.3) as *bounded below*, and satisfying (3.5.5) as *open*. If in addition to being bounded below the operator T has *closed range*, in the sense that

$$T(X) = \mathrm{cl}\, T(X) \subseteq Y , \tag{3.5.6}$$

then we shall say that T is *closed*; if instead it has dense range, in the sense that

$$\mathrm{cl}\, T(X) = X , \tag{3.5.7}$$

then we shall say that T is *dense*. Evidently (exercise) there is implication

$$\textit{left invertible} \implies \textit{closed} \implies \textit{bounded below} \implies \textit{one one} \tag{3.5.8}$$

and

$$\textit{right invertible} \implies \textit{open} \implies \textit{onto} \implies \textit{dense} ; \tag{3.5.9}$$

also

$$invitible \iff closed\ and\ dense$$
$$\iff bounded\ below\ and\ onto \iff one\ one\ and\ open \qquad (3.5.10)$$

The reader should also verify the appropriate analogues of (2.2.6)–(2.2.9) for closedness and denseness.

If X and Y are normed spaces then it is rather easy to see that the set

$$\{T \in BL(X, Y) : T\ bounded\ below\}$$

is open in the norm topology: for if $T \in BL(X, Y)$ satisfies (3.5.3) and $\|T' - T\| = k' < k$ then

$$\|T'x\| \geq \|Tx\| - \|(T' - T)x\| \geq (k - k')\|x\| .$$

We also remark that there is implication, if $T \in BL(X, Y)$ is bounded below,

$$T_n \in BL(X, Y)\ dense\ ,\ \ \|T - T_n\| \to 0 \implies T\ dense\ . \qquad (3.5.11)$$

When the normed spaces X and Y are complete then there is some simplification here, thanks to what is known as the *open mapping theorem*: it turns out that there is further implication

$$bounded\ below \implies closed\ ;\ \ onto \implies open\ . \qquad (3.5.12)$$

and hence

$$invitible \iff bounded\ below\ and\ dense \iff one\ one\ and\ onto\ . \qquad (3.5.13)$$

If X and Y are complete then by (3.5.11) a bounded below which is the limit of dense is necessarily invertible.

When X, Y and Z are complete and $T : X \to Y$ and $S : Y \to Z$ then the condition

$$S^{-1}(0) \subseteq cl\ T(X) \qquad (3.5.14)$$

gives, if $X = Y = Z$, weak exactness in the sense (1.10.1) for the pair $(S, T) \in B(X)^2$; more generally the same condition gives the analogue for the category of bounded operators between Banach spaces.

When X and Y are Banach spaces we shall write

$$BL_{00}(X, Y) = \{T \in BL(X, Y) : \dim\ T(X) < \infty\} \qquad (3.5.15)$$

for the subspace of (bounded) *finite rank* operators and

$$BL_0(X, Y) = \{T \in BL(X, Y) : \text{cl}\,\{Tx : \|x\| \le 1\}\ compact\} \tag{3.5.16}$$

for the larger subspace of *compact operators*. Evidently

$$BL_{00}(X, Y) \subseteq BL_0(X, Y) \subseteq BL(X, Y)\,; \tag{3.5.17}$$

when $Y = X$ then also $BL_{00}(X, Y)$ and $BL_0(X, Y)$ are two-sided ideals in the ring $B(X) = BL(X, X)$.

If $T \in BL(X, Y)$ and $S \in BL(Y, Z)$ then

$$S^{-1}(0) \cap T(X) = \{0\},\, S^{-1}(0) + T(X)\ closed \implies T(X)\ closed \tag{3.5.18}$$

and

$$S(Y),\, S^{-1}(0) + T(X)\ closed \implies ST(X)\ closed \implies S^{-1}(0) + T(X)\ closed\,. \tag{3.5.19}$$

The second implication of (3.5.12) is essentially the *open mapping theorem*, which depends on the Baire Category Theorem (2.4.10); a useful intermediary is *Zabreiko's Lemma* , which says that if the normed space is complete, then necessary and sufficient for a seminorm $\rho : X \to \mathbb{R}$ to be continuous is that it be *countably subadditive*, in the sense that

$$x \in \ell_1(X) \implies \rho(\sum_{n=1}^{\infty} x_n) \le \sum_{n=1}^{\infty} \rho(x_n)\,. \tag{3.5.20}$$

Of course necessary and sufficient for X to be complete is the inclusion

$$\Sigma\ell_1(X) \subseteq c_1(X)\,, \tag{3.5.21}$$

where, for each $n \in \mathbb{N}$,

$$x \in X^{\mathbb{N}} \implies (\Sigma x)_n = x_1 + x_2 \ldots + x_n\,. \tag{3.5.22}$$

A useful auxiliary concept for the proof of the open mapping theorem is that of an *almost open* operator $T \in BL(X, Y)$, for which there is $k > 0$ with the property that

$$\forall y \in Y\,,\ y \in \text{cl}\{Tx : \|x\| \le k\|y\|\}\,. \tag{3.5.23}$$

Generally if $T \in BL(X, Y)$ is bounded between normed spaces there is implication

$$T\ open \implies T\ almost\ open \implies T\ dense \tag{3.5.24}$$

and, part of (3.5.9),

$$T \ open \implies T \ onto \implies T \ dense .$$ (3.5.25)

Also, strengthening (3.5.11),

$$T \ bounded \ below \ and \ dense \implies T \ almost \ open .$$ (3.5.26)

Neither of the conditions "almost open" and "onto" implies the other; neither do "almost open" and "onto" together imply "open". If however the spaces X and Y are both complete, then all three conditions are equivalent.

3.6 Duality

If X is a normed space over $\mathbb{K} \in \{\mathbb{R}, \mathbb{C}\}$ then the space

$$X^* = BL_{\mathbb{K}}(X, \mathbb{K}) ,$$ (3.6.1)

is known as the *dual space* of X; collectively the dual space X^* acts like a somewhat disorganised "system of coordinates" for the space X, and the reason it works is due to what is known as the *Hahn-Banach theorem* which says that if $X_0 \subseteq X$ is a linear subspace then each $\varphi_0 \in (X_0)^*$ has an extension $\varphi \in X^*$ for which

$$\forall \, x \in X_0 \, : \, \varphi(x) = \varphi_0(x) ,$$ (3.6.2)

with indeed further

$$\|\varphi\| = \|\varphi_0\| .$$ (3.6.3)

The crucial consequence is that if $x, y \in X$ there is implication

$$\big(\forall \, \varphi \in X^* \, : \varphi(x) = \varphi(y)\big) \implies x = y .$$ (3.6.4)

To see this, suppose $y \neq x \in X$ and set

$$X_0 = \mathbb{C}(y - x) \ and \ \varphi_0(\lambda(y - x)) = \lambda \|y - x\| ,$$ (3.6.5)

Historically this was established first for *real* rather than complex normed spaces; the complex version was derived from the observation

$$BL_{\mathbb{R}}(X, \mathbb{R}) \cong BL_{\mathbb{C}}(X, \mathbb{C}) ,$$ (3.6.6)

valid for complex normed spaces X. Of course (3.6.6) retrospectively avoids a potential embarrassment about (3.6.1).

The word "duality" suggests a reciprocal relationship between X and X^*; but life is not quite so simple. It is clear from (3.6.4) that there is a canonical inclusion $x \mapsto x^\wedge : X \to X^{**}$ from a Banach space to its "second dual", the dual of its dual, and then by (3.6.3) this inclusion is isometric. A Banach space is said to be *reflexive* when this inclusion is onto, and hence invertible. Thus reflexive Banach spaces are isomorphic to their second dual; technically that statement is not quite what we mean: the requirement is that one particular inclusion is an "isomorphism".

Now if $T \in BL(X, Y)$ then we can define $T^* \in BL(Y^*, X^*)$ by setting

$$T^*(g) = g \circ T \ (g \in Y^*) : \tag{3.6.7}$$

evidently if $T \in BL(X, Y)$ and $S \in BL(Y, Z)$ then $(S \circ T)^* = T^* \circ S^*$ while if also $T' \in BL(X, Y)$ then $(\alpha T + \alpha' T')^* = \alpha T^* + \alpha'(T')^*$. It follows

$$T \in B(X)^{-1}_{left} \Longrightarrow T^* \in B(X^*)^{-1}_{right} ; \tag{3.6.8}$$

and

$$T \in B(X)^{-1}_{right} \Longrightarrow T^* \in B(X^*)^{-1}_{left} . \tag{3.6.9}$$

Further, if $T \in BL(X, Y)$ then

$$T \ bounded \ below \iff T^* \ almost \ open , \tag{3.6.10}$$

and dually

$$T \ almost \ open \iff T^* \ bounded \ below , \tag{3.6.11}$$

There is also implication

$$T \in BL(X, Y) \ dense \iff T^* \ one \ one , \tag{3.6.12}$$

However it is possible for $T \in BL(X, Y)$ to be one one without $T^* \in BL(Y^*, X^*)$ dense; as an example take the natural embedding

$$T = J : c_0 \subseteq \ell_\infty :$$

notice

$$(c_0)^* \cong \ell_1 ; \ (\ell_1)^* \cong \ell_\infty .$$

If we write $J_X : X \to X^{**}$ for the canonical injection $x \mapsto x^\wedge$ then it is a curiosity that the dual X^* is essentially a complemented subspace of the third dual X^{***}: specifically

$$P_X^2 = P_X = J_{X^*} \circ (J_X)^* \in B(X^{***}) \qquad (3.6.13)$$

has range

$$P_X(X^{***}) = J_{X^*}(X^*) \qquad (3.6.14)$$

and null space

$$P_X^{-1}(0) = J_{X^*}(X^*)^\perp . \qquad (3.6.15)$$

It now also follows that c_0 cannot be a dual space Y^*: for then (3.4.32) $c_0 \subsetneq \ell_\infty$ would be complemented. Linear functionals φ on the space X also combine with vectors y from the space Y to produce *rank one* operators from X to Y:

$$\varphi \odot y : x \mapsto \varphi(x)y : X \mapsto Y . \qquad (3.6.16)$$

We define $T \in BL(X, Y)$ to be *Tauberian* if

$$T^{**-1}(Y^\wedge) \subseteq X^\wedge \subseteq X^{**} . \qquad (3.6.17)$$

Evidently the space $B_{00}(X, Y)$ of (bounded) *finite rank operators* coincides with the set of finite sums of (bounded) *rank one operators*.

It is *Schauder's duality theorem* that the dual of a compact operator is compact: if $T \in B(X)$ then

$$T \in B_0(X) \iff T^* \in B_0(X^*) . \qquad (3.6.18)$$

It is not in general the case that $T \in B_0(X)$ give rise to compact multiplications $L_T : B(X) \to B(X)$ or $R_T : B(X) \to B(X)$: however

$$T \in B_0(X) \implies L_T R_T \in B_0(B(X)) , \qquad (3.6.19)$$

and also

$$T \in B_0(X) \implies L_T \wedge R_T \in B_0(\mathrm{comm}(T)) . \qquad (3.6.20)$$

Here we are writing, for R and S in $L(X, Y)$,

$$R \wedge S : (R - S)^{-1}(0) \to Y$$

for their common restriction to the subspace of X on which they agree.

In more quantitative detail with, for $T \in B(X)$,

$$\|T\|'_{ess} = \inf\{\delta > 0 : T\mathrm{Disc}(0, 1) \subseteq \mathrm{Disc}(X_\delta, \delta)\} \tag{3.6.21}$$

for some finite $X_\delta \subseteq X$, then (3.6.18)–(3.6.20) follow from

$$\|T^*\|_{ess} \leq 4\|T\|_{ess} , \tag{3.6.22}$$

$$\|L_T R_T\|'_{ess} \leq 6\|T\| \, \|T\|'_{ess} , \tag{3.6.23}$$

and

$$\|L_T \wedge R_T\|'_{ess} \leq 6\|T\|'_{ess} . \tag{3.6.24}$$

The weak* topology, on the dual E^* of a normed space E, is the topology of pointwise convergence on E, in the sense (2.5.10). The *weak topology* on a normed space E is that induced by the weak* topology of its second dual E^{**} via the embedding $x \mapsto x^\wedge$, where

$$g \in E^* \Longrightarrow x^\wedge(g) = g(x) . \tag{3.6.25}$$

Looking at the cartesian product topology of (2.5.10), we claim that a linear functional $g : E \to \mathbb{K}$ is weak* continuous if and only if

$$g \in \{x^\wedge : x \in E\} . \tag{3.6.26}$$

Note that if $(f_j)_{j \in J}$ is a finite system of linear functionals on E and $g : E \to \mathbb{K}$ is linear then (*Lemma of Auerbach*) there is, by induction on $\#(J) \in \mathbb{N}$, implication

$$\bigcap_{i \in J} f_i^{-1}(0) \subseteq g^{-1}(0) \Longrightarrow g \in \sum_{i \in J} \mathbb{K} f_i . \tag{3.6.27}$$

It is the *Krein-Milman theorem* that if E is a normed, or more generally locally convex, space, and $K \subseteq E$ is both compact and convex, then

$$K = \mathrm{cvx} \, \mathrm{ext}(K) . \tag{3.6.28}$$

This is the case for the unit ball of $E = F^*$ in its weak* topology.

Brouwer's theorem (2.7.15) also extends to infinite dimensional normed, and more generally locally convex spaces X: it is the *Schauder fixed point theorem* that if $K \subseteq X$ is compact and convex, then every continuous mapping $f : K \to K$ has a fixed point.

3.7   Enlargement

If X is a normed space we shall write, if $1 \leq p < \infty$,

$$\ell_p(X) = \{x \in X^{\mathbb{N}} : \sum_{n=0}^{\infty} \|x_n\|^p \equiv \|x\|_p^p < \infty\}\,, \tag{3.7.1}$$

in particular the bounded sequences

$$\ell_\infty(X) = \{x \in X^{\mathbb{N}} : \sup \|x_n\| < \infty\}\,, \tag{3.7.2}$$

and the null sequences

$$c_0(X) = \{x \in X^{\mathbb{N}} : \|x_n\| \to 0 \ (n \to \infty)\}\,. \tag{3.7.3}$$

Evidently $c_0(X) \subseteq \ell_\infty(X)$ is a closed subspace and now the norm on the quotient space

$$Q(X) = \ell_\infty(X)/c_0(X)\,, \tag{3.7.4}$$

is given by, for each $x \in \ell_\infty(X)$,

$$\text{dist}(x, c_0(X)) = \limsup_n \|x_n\|\,. \tag{3.7.5}$$

This is what we shall call the *enlargement* of the normed space X. Evidently $T \in BL(X, Y)$ generates an enlargement $Q(T) : Q(X) \to Q(Y)$. For us the crucial property will be the implications

$$Q(T) \ one \ one \implies T \ bounded \ below \implies Q(T) \ bounded \ below\,. \tag{3.7.6}$$

Naturally it follows that both implications are two-way. Dually, there are implications

$$Q(T) \ dense \implies T \ almost \ open \implies Q(T) \ open\,. \tag{3.7.7}$$

It is now something of an anticlimax that

$$T \in B(X)^{-1} \iff Q(T) \in B(Q(X))^{-1} : \tag{3.7.8}$$

the homomorphism Q has the Gelfand property. The enlargement of course contains as a subspace the *completion*:

$$X^{\sim} = c(X)/c_0(X)\,; \tag{3.7.9}$$

indeed

$$Q(X^{\sim}) = Q(X) . \tag{3.7.10}$$

Here $c(X)$ is the space of Cauchy sequences in X, and $c_1(X)$ the subspace of convergent sequences. Evidently

$$c_1(X) = c_0(X) + 1 \odot X$$

is the sum of the null sequences and the constant sequences.

There is a relationship between the dual of the enlargement and the enlargement of the dual: with the help of *Banach limits*

$$\varphi \in (\ell_\infty)^* \ \textit{with } c_0 \subseteq \varphi^{-1}(0) \tag{3.7.11}$$

we obtain

$$\varphi_X^\wedge : Q(X^*) \to Q(X)^*$$

given by the formula

$$\varphi_X^\wedge(f + c_0(X^*))(x + c_0(X)) = \varphi(f(x)) ;$$

with the help of this, we find implication

$$Q(T)^* \ \textit{one one} \implies Q(T^*) \ \textit{one one} . \tag{3.7.12}$$

The *bounded closure* of the range of an operator $T \in BL(X, Y)$ is given by

$$\mathrm{cl}^{\sim}(T, X) = \{\lim_n T x_n : x \in \ell_\infty(X) \cap T^{-1} c_1(Y)\} . \tag{3.7.13}$$

Evidently

$$T(X) \subseteq \mathrm{cl}^{\sim}(T, X) \subseteq \mathrm{cl}(TX) ; \tag{3.7.14}$$

Also, with $\mathbf{q}(y) = y + c_0(Y) \in Q(Y)$,

$$\mathrm{cl}^{\sim}(T, X) = \{y \in Y : \mathbf{q}(y) \in Q(T)Q(X)\} . \tag{3.7.15}$$

The idea is latent in an old proof of Dieudonné that, if $K \in B_0(X) \subseteq B(X)$ is compact in the sense (3.5.16), the operator $T = I + K$ has closed range.

If $T \in BL(X, Y)$, between Banach spaces X and Y, and $M \subseteq T(X) \subseteq Y$ is a linear subspace, then there is equivalence

$$\mathrm{cl}\, M \subseteq T(X) \iff \ell_\infty(M) \subseteq T(\ell_\infty(X)) \iff \ell_\infty(M) \subseteq T(X) + c_0(Y) .$$
$$(3.7.16)$$

Here we write T for the induced operator $\ell_\infty(T)$: $\ell_\infty(X) \rightarrow \ell_\infty(Y)$, and sometimes $y = Tx \in \ell_\infty(Y)$ for the obvious constant sequence. There is forward implication in (3.7.16) when X is complete, and backward when Y is complete. Whether or not either space is complete, the right hand inclusion holds if and only if the quotient operator

$$T_M^\wedge : T^{-1}(M)/T^{-1}(0) \rightarrow Y$$

is bounded below.

For example, recalling (1.3.14), a normed algebra element $a \in A$ will be called *almost regular* provided

$$a \in \mathrm{cl}^\sim(L_a R_a, A) .$$
$$(3.7.17)$$

When the normed algebra A is complete, there is implication, if $a \in A$,

$$a \text{ almost regular} \implies a \text{ regular} .$$
$$(3.7.18)$$

There is also an *essential enlargement*

$$\mathbf{P}(X) = \ell_\infty(X)/m(X) ,$$
$$(3.7.19)$$

where

$$m(X) = \{x \in \ell_\infty(X) : \mathrm{cl}\{x_n : n \in \mathbb{N}\} \text{ compact}\} .$$
$$(3.7.20)$$

If $x \in \ell_\infty(X)$ then (cf (2.5.8))

$$\mathrm{dist}(x, m(X)) = \inf\{\delta > 0 : \{x_n : n \in \mathbb{N}\} \text{ has a finite } \delta \text{ net}\}$$
$$(3.7.21)$$

is a "measure of non compactness" for the range of x.

If $T \in BL(X, Y)$ then there is inclusion

$$Tm(X) \subseteq m(Y) ,$$
$$(3.7.22)$$

and implication

$$T\ell_\infty(X) \subseteq m(Y) \iff T \in BL_0(X, Y) .$$
$$(3.7.23)$$

Analagous to (3.7.6) there is implication, when X and Y are complete,

$$\mathbf{P}(T) \ one \ one \implies T \ upper \ semi \ Fredholm \implies \mathbf{P}(T) \ bounded \ below \ .$$
(3.7.24)

Dually

$$\mathbf{P}(T) \ dense \implies T \ lower \ semi \ Fredholm \implies \mathbf{P}(T) \ open \ . \quad (3.7.25)$$

Obviously therefore

$$\mathbf{P}(T) \ invertible \iff T \ Fredholm \ . \quad (3.7.26)$$

Here, on Banach spaces, "upper semi Fredholm", the essential version of "bounded below", means closed range and finite dimensional null space, while "lower semi Fredholm", the essential version of onto, means closed range of finite codimension.

Notice the two-way implication

$$m(X) = \ell_\infty(X) \iff \dim(X) < \infty \ . \quad (3.7.27)$$

3.8 Functions

If X is a set then the mappings $A = \mathbb{C}^X = \text{Map}(X, \mathbb{C})$ form a linear algebra, with

$$a \in A^{-1} \iff 0 \notin a(X) = \{a(t) : t \in X\} \ . \quad (3.8.1)$$

This is easily verified: if $a(t)$ never vanishes then we can successfully define $a^{-1} \in A$ by setting

$$a^{-1}(t) = a(t)^{-1} \ (t \in X) \ . \quad (3.8.2)$$

Conversely if there exists b for which $ba = 1$ then $b(t)a(t) = 1$ for all $t \in X$, preventing $a(t) = 0$.

Since A is commutative there can be no distinction between left, right and two-sided invertibility. The same is true of the "monomorphisms" and "epimorphisms" A^o; also here there is equality

$$A^{-1} = A^o_{left} = A^o_{right} \ . \quad (3.8.3)$$

Inclusion one way round is (1.3.3); conversely if ever $a(t) = 0$ then with $b = \delta_t$, where

$$\delta_t(s) = 1 \ (s = t) \ , \ = 0 \ (s \neq t) \quad (3.8.4)$$

it is clear $ab = ba = 0 \neq b$. We shall also, more generally, define the *characteristic function*, of $K \subseteq X$, by setting

$$\delta_K(t) = 1 \ (t \in K) \ , \quad \delta_K(t) = 0 \ (t \in X \setminus K) \ . \tag{3.8.5}$$

If instead we consider the algebra

$$A = \ell_\infty(X) = \{a \in \mathbb{C}^X : \sup\{|a(t)| : t \in X\} < \infty\} \tag{3.8.6}$$

then we still have

$$A^o_{left} = A^o_{right} = \{a \in A : 0 \notin a(X)\} \supseteq A^{-1} \ , \tag{3.8.7}$$

with equality at the end if and only if the set X is finite. More interesting is the case of a topological space X and the algebra $A = C(X)$ of continuous functions: for the "normal" spaces of (2.4.14), Urysohn's Lemma enables us to see

$$A^{-1} = \{a \in A : 0 \notin a(X)\} \ . \tag{3.8.8}$$

We lose even this for the algebra $A = C_\infty(X) = C(X) \cap \ell_\infty(X)$ of bounded continuous functions: but in the crucially important case in which the topological space X is compact and separated then it turns out that

$$C(X) = C_\infty(X) \ , \tag{3.8.9}$$

The good news is that here we again get equality (3.8.3). We remark that for arbitrary X the algebra $\ell_\infty(X)$, and for compact separated X the algebra $C(X)$, are Banach algebras, with

$$\|a\| = \sup_{t \in X} |a(t)| \ ; \tag{3.8.10}$$

one of the triumphs of spectral theory has been to essentially reduce the study of all commutative Banach algebras to $C(X)$.

3.9 Topological Zero Divisors

The "zero divisors" in a topological ring lie inside the larger sets of *topological zero divisors*, for which the left or right multiplication operators fail to be bounded below.

If A is a normed algebra we shall write

$$A^{\bullet}_{left} = \{a \in A : L_a \ bounded \ below\} \ , \tag{3.9.1}$$

and

$$A^\bullet_{right} = \{a \in A : R_a \text{ bounded below}\} \tag{3.9.2}$$

for the *strong monomorphisms* and *strong epimorphisms*. Evidently

$$A^{-1}_{left} \subseteq A^\bullet_{left} \subseteq A^o_{left} . \tag{3.9.3}$$

and

$$A^{-1}_{right} \subseteq A^\bullet_{right} \subseteq A^o_{right} . \tag{3.9.4}$$

We also have implication

$$x, y \in A^\bullet_{left} \implies xy \in A^\bullet_{left} \implies y \in A^\bullet_{left} , \tag{3.9.5}$$

and

$$x, y \in A^\bullet_{right} \implies xy \in A^\bullet_{right} \implies x \in A^\bullet_{right} . \tag{3.9.6}$$

Both $A^\bullet_{left}$ and $A^\bullet_{right}$ are open subsets of A. We notice also, recalling (3.5.11),

$$\mathrm{cl}(A^{-1}) \cap A^\bullet_{left} = A^{-1} = \mathrm{cl}(A^{-1}) \cap A^\bullet_{right} . \tag{3.9.7}$$

If the homomorphism $T : A \to B$ is invertible then obviously $T(A^{-1}) = B^{-1}$ and $T^{-1}(B^{-1}) = A^{-1}$. More generally, if $T : A \to B$ is left invertible, with a multiplicative left inverse, then $T^{-1}(B^{-1}_{left}) \subseteq A^{-1}_{left}$, and similarly if T is right invertible.

If the homomorphism $T : A \to B$ is bounded below then there is inclusion

$$T^{-1} B^\bullet_{left} \subseteq A^\bullet_{left} ; \; T^{-1} B^\bullet_{right} \subseteq A^\bullet_{right} , \tag{3.9.8}$$

while if $T : A \to B$ is one one there is inclusion

$$T^{-1} B^o_{left} \subseteq A^o_{left} ; \; T^{-1} B^o_{right} \subseteq A^o_{right} . \tag{3.9.9}$$

If A is a normed algebra, or more generally a *normed linear category*, then intermediate between weak exactness and splitting exactness for the pair $(b, a) \in A^2$ would be the condition of *normed exactness*, that there are $k > 0$ and $h > 0$ for which, for arbitrary $(u, v) \in A^2$,

$$\|vu\| \leq k\|bu\| \, \|v\| + h\|u\| \, \|va\| . \tag{3.9.10}$$

When in particular $A = B(X)$ it is equivalent that for arbitrary $x \in X$ and $\varphi \in X^*$

$$|\varphi(x)| \le k\|bx\| \, \|\varphi\| + h\|x\| \, \|\varphi a\| . \qquad (3.9.11)$$

Evidently, in the sense (3.9.10),

$$(b, 0) \; normed \; exact \iff b \in A^{\bullet}_{left} \qquad (3.9.12)$$

while

$$(0, a) \; normed \; exact \iff a \in A^{\bullet}_{right} . \qquad (3.9.13)$$

For example the *quasinilpotents* of (3.3.8) are both left and right topological zero divisors.

3.10 The Riesz Lemmas

We take here the opportunity to group together three rather technical results connecting a subspace and a point. If $Y \subseteq X$ is a subspace of a normed space X and $x \in X$ then there is a sequence $y \in Y^{\mathbb{N}}$ for which

$$\|y_n - x\| \to \mathrm{dist}(x, Y) \; with \; \|y_n\| \le \|x\| + \mathrm{dist}(x, Y) \le 2\|x\| . \qquad (3.10.1)$$

Also if $x \notin Y$ then for each $t \in [0, 1) \subseteq \mathbb{R}$ there is $x_t \in X$ for which

$$\mathrm{cl}(Y) + \mathbb{K}x_t = \mathrm{cl}(Y) + \mathbb{K}x \; with \; t\|x_t\| \le \mathrm{dist}(x, Y) . \qquad (3.10.2)$$

If $x \notin \mathrm{cl}(Y)$ then there is $M > 0$ for which, for arbitrary $y \in Y$ and $\lambda \in \mathbb{K}$,

$$\|y + \lambda x\| \le \|y\| + |\lambda| \, \|x\| \le M\|y + \lambda x\| . \qquad (3.10.3)$$

Traditionally the term "Riesz Lemma" refers to (3.10.2), which enables us to nearly "drop a perpendicular" from x to Y, and shows that if X is infinite dimensional its unit ball cannot be (norm) compact. (3.10.1) is one of the foundations of "gap theory"; (3.10.3) says that all norms on X which agree on Y give the same topology on $Y + \mathbb{K}x$. It follows that a finite dimensional subspace $Y \subseteq X$ has a unique norm topology, and is always complete, hence also closed in the ambient space Y. It is, further, always *complemented*, in the sense (3.4.21); equivalently

$$Y = P(X) \; with \; P = P^2 \in B(X) . \qquad (3.10.4)$$

This comes from the *Lemma of Auerbach* which (together with the Hahn-Banach theorem), says that if

$$e \in X^k \text{ is linearly independent with } \|e_j\| = 1 \ (j = 1, 2, \ldots, k) \tag{3.10.5}$$

then there is

$$f \in (X^*)^k \text{ with } \|f_i\| = 1 \ (i = 1, 2, \ldots, k) . \tag{3.10.6}$$

for which

$$f_i(e_j) = \delta_{ij} \ (\{i, j\} \subseteq \{1, 2, \ldots, k\}) . \tag{3.10.7}$$

Thus (3.10.4) holds with

$$Y = \sum_{j=1}^{k} \mathbb{K}e_j \ , \quad P = \sum_{j=1}^{k} f_j \odot e_j . \tag{3.10.8}$$

This gives the implication (3.6.27), with $J = \{1, 2, \ldots, k\}$,

$$\bigcap_{i \in J} f_i^{-1}(0) = P^{-1}(0) = (I - P)(X) ; \ \subseteq g^{-1}(0) \iff g = g \circ P = \sum_{j=1}^{k} g(e_j)f_j . \tag{3.10.9}$$

A proof of (3.10.7), by induction on k, uses the Hahn-Banach theorem and (3.10.3).

Equation (3.10.1) also says that, in the notation of (3.7.15), that if $T = J_Y : Y \to X$ is the embedding, then $\text{cl}(Y) = \text{cl}^{\sim}(T, Y)$.

If $Y \subseteq X$ and $Z \subseteq X$ are closed subspaces then the *gap* between Y and Z is given by

$$\text{gap}(Y, Z) = \max(\delta(Y, Z), \delta(Z, Y)) , \tag{3.10.10}$$

where

$$\delta(Y, Z) = \sup\{\text{dist}(x, Z) : x \in Y , \ \|x\| \le 1\} \tag{3.10.11}$$

If $J_Y : Y \to X$ and $K_Y : X \to X/Y$ are the natural embedding and quotient then

$$\delta(Y, Z) = \|K_Z J_Y\| . \tag{3.10.12}$$

For arbitrary $x \in X$ we have, using (3.10.1), inequality

$$\text{dist}(x, Y) \le \delta(Y, Z)\|x\| + (1 + \delta(Y, Z))\text{dist}(x, Z) . \tag{3.10.13}$$

More generally, if $T \in B(X)$ and $x \in X$,

$$\|Tx\| \leq \|T\|\mathrm{dist}(x, Y) + \|T_Y\|(\|x\| + \mathrm{dist}(x, Y)) . \tag{3.10.14}$$

The *angle* between Y and Z is given by

$$\gamma(Y, Z) = \inf\{\mathrm{dist}(x, Z) : x \in Y , \ \|x\| \geq 1\} . \tag{3.10.15}$$

Anderson and Foias have shown that there is implication

$$\gamma(Y, Z) = 0 \Longleftrightarrow \gamma(Z, Y) = 0 . \tag{3.10.16}$$

to see this observe that, whether or not $\gamma(Y, Z) < 1$,

$$(1 - \gamma(Y, Z))\gamma(Z, Y) \leq \gamma(Y, Z) . \tag{3.10.17}$$

The operator $K_Z J_Y$ of (3.10.12) is a vital component of Yang's "one diagram proof" of the index theorem (1.9.8).

3.11 Polar Decomposition

A *generalized polar decomposition* in a C* algebra A is a pair $(u, c) \in A^2$ for which

$$u = uu^*u ; \tag{3.11.1}$$

$$c = c^* \in \mathrm{Re}(A) ; \tag{3.11.2}$$

$$L_a^{-1}(0) \subseteq L_c^{-1}(0) ; \tag{3.11.3}$$

When

$$a = uc \in A$$

then (u, c) is said to be a generalized polar decomposition for a. The condition (3.11.1) characterizes the *partial isometries* $u \in A$. If $(u, c) \in A^2$ is a generalized polar decomposition for $a \in A$ then

$$a^*a = c^2 \text{ and } u^*a = c , \tag{3.11.4}$$

and also

$$aa^*u = ua^*a , \tag{3.11.5}$$

It is clear from (3.11.4) and (3.11.5) that

$$\{u, c\} \subseteq \mathrm{comm}^2(a, a^*) \, . \tag{3.11.6}$$

If in addition

$$0 \leq c \in A^+ \text{ and } L_c^{-1}(0) \subseteq L_u^{-1}(a) \, , \tag{3.11.7}$$

giving equality in (3.11.3), then $(u, c) \in A^2$ is called a *polar decomposition*.

The polar decomposition of $a \in A$, when it exists, is uniquely determined, and we shall write

$$(u, c) = (\mathrm{sgn}(a), |a|) \, . \tag{3.11.8}$$

with cf (3.4.6) necessarily

$$|a| = (a^*a)^{1/2} \, . \tag{3.11.9}$$

Evidently

$$\mathrm{sgn}(a^*) = \mathrm{sgn}(a)^* \, ; \ |a^*| = |a| \, . \tag{3.11.10}$$

If $a \in A$ has a Moore-Penrose inverse $a^\dagger$ then it also has a polar decomposition:

$$a \in A^\dagger \Longrightarrow \mathrm{sgn}(a) = (a^\dagger)^*|a| = (a^*)^\dagger|a| \, . \tag{3.11.11}$$

When $A = B(X)$ for a Hilbert space X then every $a \in A$ has a polar decomposition: if $a \in A$ and $c = |a| = (a^*a)^{1/2}$ then define $u = \mathrm{sgn}(a)$ by setting, for arbitrary $y \in \mathrm{cl}\, a(X)$ and $z \in a(X)^\perp$,

$$u(y + z) = \lim_{cx_n \to y} a(x_n) \, . \tag{3.11.12}$$

At the other extreme if for example $A = C[0, 1] \equiv C([0, 1])$ then only $a = 0$ and the invertibles $a \in A^{-1}$ have polar decompositions. In general, even for $A = B(X)$ with two-dimensional X, the "modulus" $a \mapsto |a|$ of (3.11.9) fails to satisfy the triangle inequality. The partial isometries of (3.11.1) include the *unitaries*, which satisfy

$$u^*u = 1 = uu^* \, , \tag{3.11.13}$$

which are of course the invertible partial isometries. When $A = B(X)$, for a Hilbert space X, the unitary group is topologically contractible, in the sense of Sect. 2.7. This observation, together with the polar decomposition, is the essence of Kuiper's theorem, and (4.6.14) below. Generally if $a = uc \in A$ has polar decomposition

(u, c) then $c + 1 - u^*u \in A^o_{left}$ is monomorphic, in the sense (1.3.2), and necessary and sufficient for $a \in A^\cap$ is that

$$c + 1 - u^*u \in A^{-1} \, ; \tag{3.11.14}$$

thus

$$a \in A^\cap \Longleftrightarrow |a| + 1 - \text{sgn}(a)^*\text{sgn}(a) \in A^\cap . \tag{3.11.15}$$

Necessary and sufficient for $a \in A^\cup$ to be decomposably regular is that it "has index zero", in the sense that its support and co-support projections are "Murray von Neumann equivalent", so that there is $w \in A$ for which

$$1 - u^*u = w^*w \text{ and } 1 - uu^* = ww^* . \tag{3.11.16}$$

We remark also that, for a C* algebra A, every regular element $a \in A$ has a Moore-Penrose inverse: there is equality

$$A^\cap = A^\dagger . \tag{3.11.17}$$

Specifically, if $a = aba \in A^\cap$ then with $ba = p = p^2$ and $ab = q = q^2$ then also

$$e = [p] = (p^*p(1 - (p - p^*)^2))^{-1} = e^2 \tag{3.11.18}$$

and

$$f = [q^*] = (qq^*(1 - (q - q^*)^2))^{-1} = f^2 . \tag{3.11.19}$$

and then

$$a^\dagger = ebf . \tag{3.11.20}$$

This is sometimes known as the "poor man's path", between p and q for which $1 - (p - q)^2 \in A^{-1}$. It also true in C*algebras A that

$$L_a(A) = \text{cl } L_a(A) \Longrightarrow a \in A^\cap \tag{3.11.21}$$

Spectral Theory

4

The colours of the rainbow are known as the "visible spectrum", and chemical elements can sometimes be identified by analysing how much of this is retained when they interact with light. In the rarified atmosphere of linear algebras there is something similar.

4.1 Spectrum

If A is a complex linear algebra, with identity 1 and invertible group A^{-1}, then to each element $a \in A$ corresponds its *spectrum*,

$$\sigma(a) \equiv \sigma_A(a) = \{\lambda \in \mathbb{C} : a - \lambda \notin A^{-1}\}. \tag{4.1.1}$$

For example if $A = \mathbb{C}$ then each element constitutes its own spectrum:

$$\lambda \in \mathbb{C} \Longrightarrow \sigma(\lambda) = \{\lambda\}. \tag{4.1.2}$$

Readers will probably have encountered the spectrum of a *matrix*:

$$a \in \mathbb{C}^{n \times n} \Longrightarrow \sigma(a) = \{\lambda \in \mathbb{C} : \det(a - \lambda) = 0\}, \tag{4.1.3}$$

the roots of the *Cayley-Hamilton polynomial*. Much more revealing is (3.8.1) the spectrum of a *function*:

$$a \in \mathbb{C}^X \equiv \mathrm{Map}(X, \mathbb{C}) \Longrightarrow \sigma(a) = \{a(t) : t \in X\}. \tag{4.1.4}$$

In words, the spectrum of a function is its *range*. This suggests how much, and how little, the spectrum is able to say about an algebra element.

In this purely algebraic environment it is possible that the spectrum of $a \in A$ is empty, or unbounded, or indeed the whole of $\mathbb{C}$. It does however cooperate with *polynomials*: if $a \in A$ and $p \in$ Poly then there is equality

$$\sigma p(a) = p\sigma(a) , \qquad (4.1.5)$$

unless

$$p = p(0) \text{ is constant and } \sigma(a) = \emptyset \text{ is empty} . \qquad (4.1.6)$$

Here $p(a)$ is given by (1.12.8). Equality (4.1.5) breaks into two equal and opposite inclusions, one of which is much easier than the other. Inclusion

$$p\sigma(a) \subseteq \sigma p(a) \qquad (4.1.7)$$

is a consequence of the *remainder theorem* for polynomials: if $\lambda \in \mathbb{C}$ then there is a polynomial $q \in$ Poly for which

$$p - p(\lambda) = (z - \lambda)q \in \text{Poly} ,$$

and hence

$$p(a) - p(\lambda) = (a - \lambda)q(a) = q(a)(a - \lambda) \in A .$$

It follows

$$p(a) - p(\lambda) \in A^{-1} \Longrightarrow a - \lambda \in A^{-1} ,$$

giving indeed (4.1.7). The opposite inclusion,

$$\sigma p(a) \subseteq p\sigma(a) , \qquad (4.1.8)$$

is much deeper, and uses the fundamental theorem of algebra (2.7.20): if $\mu \in \mathbb{C}$ then, unless the polynomial p is constant, there is $0 \neq \alpha \in \mathbb{C}$ giving equality

$$p - \mu = \alpha \prod \{(z - \lambda)^{\nu(\lambda)} : \lambda \in p^{-1}(\mu)\} , \qquad (4.1.9)$$

Now—if necessary by induction on the cardinal number $\#p^{-1}(\mu)$—there is implication

$$(p(\lambda) = \mu \Longrightarrow a - \lambda \in A^{-1}) \Longrightarrow p(a) - \mu \in A^{-1} ,$$

giving (4.1.8) for non constant p. If $p = \mu$ is constant then $\mu = p(\lambda)$ for arbitrary $\lambda \in \mathbb{C}$.

The reader may like to extend (4.1.5) to rational functions p/q: she will notice that—thanks to (4.1.5)—

$$\exists (p/q)(a) \iff \sigma(a) \cap q^{-1}(0) = \sigma(a) \cap \sigma(p/q) = \emptyset .\qquad (4.1.10)$$

It is at this point in the discussion that the importance of complex linear algebras shows itself: the fundamental theorem of algebra is not available in the real field.

Naturally inclusion (4.1.8) will fail if p is constant and $\sigma(a) = \emptyset$, unless of course $A = \{0\}$ is the singleton.

4.2 Left and Right Point Spectrum

The invertible group $A^{-1} \subseteq A$ for a linear algebra is (1.2.7) the intersection of the left and the right invertibles, and hence if $a \in A$

$$\sigma(a) = \sigma^{left}(a) \cup \sigma^{right}(a) ,\qquad (4.2.1)$$

where

$$\sigma^{left}(a) = \{\lambda \in \mathbb{C} : a - \lambda \notin A^{-1}_{left}\} , \ \ \sigma^{right}(a) = \{\lambda \in \mathbb{C} : a - \lambda \notin A^{-1}_{right}\} .$$
$$(4.2.2)$$

Thanks to the implications (1.2.4) and (1.2.5), the spectral mapping property (4.1.7) holds separately for the left and the right spectrum: the argument is the same. The left and right zero divisors give rise to two further subsets,

$$\pi^{left}(a) = \{\lambda \in \mathbb{C} : a - \lambda \notin A^{o}_{left}\} , \ \ \pi^{right}(a) = \{\lambda \in \mathbb{C} : a - \lambda \notin A^{o}_{right}\} ,$$
$$(4.2.3)$$

recalling the "monomorphisms" and "epimorphisms" of (1.3.2). By (1.3.3) there is inclusion

$$\pi^{left}(a) \subseteq \sigma^{left}(a) , \ \ \pi^{right}(a) \subseteq \sigma^{right}(a) ,\qquad (4.2.4)$$

but by (1.3.4) also, improving (4.2.1),

$$\sigma(a) = \sigma^{left}(a) \cup \pi^{right}(a) = \pi^{left}(a) \cup \sigma^{right}(a) .\qquad (4.2.5)$$

The one-way spectral mapping theorem (4.1.7) holds for the left and the right point spectrum, thanks to the implications (1.3.5) and (1.3.6).

If for example $A = L(X)$ is the linear operators on the vector space X then by (1.13.7) and (1.13.8) there is equality, for each $T \in L(X)$,

$$\sigma^{left}(T) = \pi^{left}(T) , \ \sigma^{right}(T) = \pi^{right}(T) . \tag{4.2.6}$$

If instead $A = \mathbb{C}^X$ is the functions on the set X then, for each $f : X \to \mathbb{C}$ and each polynomial p,

$$\sigma(f) = \pi^{left}(f) = \pi^{right}(f) = f(X) , \tag{4.2.7}$$

and equality

$$p\sigma(f) = \{p(f(t)) : t \in X\} = (p \circ f)(X) = \sigma p(f)$$

is visible.

In a general algebra A there is equality, by (1.13.9) and (1.13.10),

$$\pi^{left}(a) = \pi^{left}(L_a) , \ \pi^{right}(a) = \pi^{left}(R_a) . \tag{4.2.8}$$

We claim that also

$$\sigma^{left}(a) = \sigma^{left}(L_a) = \sigma^{right}(R_a) \tag{4.2.9}$$

and

$$\sigma^{right}(a) = \sigma^{right}(L_a) = \sigma^{left}(R_a) , \tag{4.2.10}$$

and hence

$$\sigma(a) = \sigma(L_a) = \sigma(R_a) . \tag{4.2.11}$$

Generally if $T : A \to B$ is a homomorphism there is inclusion

$$\sigma(Ta) \equiv \sigma_B(Ta) \subseteq \sigma_A(a) \equiv \sigma(a) , \tag{4.2.12}$$

and similarly for the left and the right spectrum: this is (1.5.3) and (1.5.4). Conversely if $T : A \to B$ has the Gelfand property (1.5.6) then there is equality throughout (4.2.12), giving indeed *spectral permanence*

$$\sigma_A(a) = \sigma_B(Ta) . \tag{4.2.13}$$

It follows from (4.2.12) that the second term in each of (4.2.9) and (4.2.10) is included in the first, and by similar argument, involving anti homomorphisms, so is the third. The reverse inclusions are a combination of (4.2.6) and (4.2.8).

4.3   Banach Algebra Spectral Theory

Spectral theory comes into its own in Banach algebras, where the spectrum is closed
and bounded, and very very nearly always nonempty. The easier bits of this flow
from the inoffensive (3.4.1): the invertible group is open and it is clear by the
continuity of the *resolvent function*

$$(z - a)^{-1} : \mathbb{C} \setminus \sigma(a) \to A \qquad (4.3.1)$$

that the complement of the spectrum

$$\mathbb{C} \setminus \sigma(a) = (z - a)^{-1}((a + A^{-1}) \cap \mathbb{C}) \qquad (4.3.2)$$

is an open subset of $\mathbb{C}$. Also by (3.4.1)

$$|\lambda| > \|a\| \implies \lambda - a = \lambda(1 - \lambda^{-1}a) \in A^{-1} \iff \lambda \notin \sigma(a)\,,$$

so that

$$\sigma(a) \subseteq \|a\|\mathbb{D} \equiv \{\lambda \in \mathbb{C} : |\lambda| \le \|a\|\}\,. \qquad (4.3.3)$$

Here $\mathbb{D} \subseteq \mathbb{C}$ is the closed unit disc. This tells us that the spectrum in a Banach
algebra is always closed and bounded, therefore *compact*. To see why it is nonempty
one needs to reach into complex analysis for *Liouville's theorem*, which says that

if $f : \mathbb{C} \to \mathbb{C}$ is holomorphic and bounded then $f = f(0)$ is constant.   (4.3.4)

This extends to holomorphic functions from $\mathbb{C}$ to Banach spaces, in particular
Banach algebras, and applies to the resolvent function (4.3.1): thus if indeed the
spectrum were empty then the resolvent would be everywhere defined. By the
second part of (3.4.1) it is also clear that

$$\|(\lambda - a)^{-1}\| \to 0 \ (|\lambda| \to \infty)\,; \qquad (4.3.5)$$

this is enough not only to guarantee that it would therefore be bounded, but also to
insist that if it were constant then that constant would have to be zero. It thus follows
that if $a \in A$ then there is implication

$$\sigma(a) = \emptyset \implies 1 = 0 \implies A = \{0\}\,. \qquad (4.3.6)$$

Liouville's theorem is of course in the same ball park as the fundamental theorem
of algebra (2.7.20), and can be derived from the *Cauchy integral formula*: if $a \in A$
and $f : U \to \mathbb{C}$ is holomorphic on an open set $\sigma(A) \subseteq U \subseteq \mathbb{C}$ then we can

successfully define

$$f(a) = \frac{1}{2\pi i} \oint_{\sigma(a)} f(z)(z-a)^{-1} dz \,. \tag{4.3.7}$$

This in turn gives the *maximum modulus principle*: if $R > 0$ and $\sigma(a) \subseteq \text{int } R\mathbb{D}$ then

$$K = R\mathbb{D} \implies \sup_K \|f(\cdot)\| \le \sup_{\partial K} \|f(\cdot)\| \,. \tag{4.3.8}$$

Differentiating the Cauchy integral formula and allowing $R \to \infty$ shows that the derivative df/dz vanishes outside $\eta\sigma(a)$. Liouville's theorem, applied to the polynomial reciprocal $1/p$, gives back the fundamental theorem of algebra (2.7.20): for if $p^{-1}(0) = \emptyset$ then (4.3.4) and (4.3.5) would say $1/p \equiv 0$. The same source gives two different expressions for the *spectral radius* of $a \in A$:

$$|a|_\sigma \equiv \sup\{|\lambda| : \lambda \in \sigma(a)\} = \lim_{n \to \infty} \|a^n\|^{1/n} \,. \tag{4.3.9}$$

From (4.3.9) it follows that in a Banach algebra the quasinilpotents can be characterized by their spectrum:

$$a \in \text{QN}(A) \iff \sigma(a) \subseteq \{0\} \,. \tag{4.3.10}$$

Another characterization is now evident:

$$\text{QN}(A) = \{a \in A : 1 - \text{comm}(a)a \subseteq A^{-1}\} \,. \tag{4.3.11}$$

The quasinilpotents offer the ultimate generalization of "simply polar": we shall say that $a \in \text{QP}(A) \subseteq A$ is *quasipolar* if it has a spectral projection $a^\bullet = q \in A$ for which

$$q^2 = q \,; \ aq = qa \,; \ a + q \in A^{-1} \,; \ aq \in \text{QN}(A) \,. \tag{4.3.12}$$

When $a \in \text{QP}(A)$ is quasipolar then it has a *Koliha-Drazin inverse*

$$a^\times = (a+q)^{-1}(1 - a^\bullet) \,. \tag{4.3.13}$$

It turns out that $a^\bullet$ is uniquely determined and double commutes with a. The quasipolars can also be identified spectrally:

$$a \in \text{QP}(A) \iff 0 \notin \text{acc } \sigma(a) \,. \tag{4.3.14}$$

Generally if $T : A \to B$ is a Banach algebra homomorphism there is inclusion

$$T \, \text{QP}(A) \subseteq \text{QP}(B) \tag{4.3.15}$$

and hence

$$QP(A) \subseteq T^{-1}QP(B) : \qquad (4.3.16)$$

equality here will be described as *Drazin permanence*. From (4.3.14) it is clear, whether or not the homomorphism $T : A \to B$ is one one, that there is implication

$$\textit{spectral permanence implies Drazin permanence .} \qquad (4.3.17)$$

Also, however

$$\textit{Drazin permanence and one one together imply simple permanence .} \qquad (4.3.18)$$

It follows, completing the circle of (1.8.14) and (1.8.15), that for Banach algebra homomorphisms, spectral permanence, simple permanence and one one make up a "democratic consensus" in the sense (1.2.17).

Generally, in a (complex) linear algebra we shall call an element $a \in A$ a *rank one element* if there is inclusion

$$aAa \subseteq \mathbb{C}a , \qquad (4.3.19)$$

and hence a linear functional $\tau_a : A \to \mathbb{C}$ for which

$$L_a \circ R_a = \tau_a \odot a ; \qquad (4.3.20)$$

for normed algebras we ask that $\tau_a \in A^*$ is bounded. In Banach algebras, the rank one elements can be characterized spectrally: necessary and sufficient is implication

$$x \in A \implies \#\sigma(xa \setminus \{0\}) \leq 1 . \qquad (4.3.21)$$

Finite sums of rank one elements make up two sided ideal Soc(A), known as the Socle . We remark that a semi simple Banach algebra is finite dimensional if, and only if, every invertible element has a finite spectrum, and is of dimension 1 if the spectrum is always a singleton.

In a C* algebra A, self-adjoint elements have real spectrum:

$$a = a^* \in A \implies \sigma(a) \subseteq \mathbb{R} . \qquad (4.3.22)$$

To see this suppose $0 \neq t \in \mathbb{R}$ and argue that, for arbitrary $s, r \in \mathbb{R}$ there is implication,

$$(r+t)^2 > \|a+s\|^2 + r^2 = \|a+s-ir\|^2 \implies a+s+it = i(r+t)(1+\frac{a+s-ir}{i(r+t)}) \in A^{-1} ; \qquad (4.3.23)$$

to satisfy the condition on the left hand side set for example $2tr = \|a\|^2$.

This motivates the introduction of the *positive cone* of a C* algebra A:

$$A^+ = \{a \in A : a = a^* \text{ and } \sigma(a) \subseteq [0, \infty) = \mathbb{R}^+ \subseteq \mathbb{C}\}. \tag{4.3.24}$$

We shall also write, for the invertible positives,

$$A^{++} = A^+ \cap A^{-1}. \tag{4.3.25}$$

The "square root lemma" (3.4.6) can be sharpened in a C* algebra: if $a = a^*$ and $\|a\| < 1$ then (3.4.6) binomial expansion gives the uniquely determined $b \in A^+$ for which $b^2 = 1 - a$. It follows that, for a C* algebra A,

$$A^+ = \{a^*a : a \in A\} = \{c^2 : c = c^* \in A\}. \tag{4.3.26}$$

If $a = a^* \in A$ and $\|a\| < k \in \mathbb{R}$ then the following are equivalent:

$$0 < t \in \mathbb{R} \Longrightarrow a + t \in A^{-1}; \tag{4.3.27}$$

$$\|k - a\| < k. \tag{4.3.28}$$

Hence there is inclusion

$$A^+ + A^+ \subseteq A^+ \tag{4.3.29}$$

and if $\{a, b\} \subseteq A^+$ then there is topological norm additivity".

$$\text{Max}(\|a\|, \|b\|) \leq \|a + b\| \leq \|a\| + \|b\|. \tag{4.3.30}$$

4.4　Approximate Point Spectrum

Topological zero divisors give rise to "approximate point spectrum":

$$\tau^{left}(a) = \{\lambda \in \mathbb{C} : a - \lambda \notin A^\bullet_{left}\} \tag{4.4.1}$$

and

$$\tau^{right}(a) = \{\lambda \in \mathbb{C} : a - \lambda \notin A^\bullet_{right}\}. \tag{4.4.2}$$

Evidently

$$\pi^{left}(a) \subseteq \tau^{left}(a) \subseteq \sigma^{left}(a) \tag{4.4.3}$$

and

$$\pi^{right}(a) \subseteq \tau^{right}(a) \subseteq \sigma^{right}(a) . \tag{4.4.4}$$

By essentially the same arguments as (1.11.5) and (1.11.6) it is clear that

$$\tau^{left}(a) = \tau^{left}(L_a) , \ \tau^{right}(a) = \tau^{left}(R_a) ; \tag{4.4.5}$$

in the algebra $B(X)$ there is equivalence

$$T \ bounded \ below \ on \ X \iff L_T \ bounded \ below \ on \ B(X) . \tag{4.4.6}$$

It turns out, with a little duality theory, that with $T \in A = B(X)$,

$$\tau^{right}(T) = \{\lambda \in \mathbb{C} : (T - \lambda I)(X) \neq X\} : \tag{4.4.7}$$

the "right approximate point spectrum" is in a sense the "onto spectrum". This in turn has consequences in general Banach algebras: if $a \in A$ then

$$\tau^{right}(L_a) = \sigma^{right}(L_a) = \sigma^{right}(a) \tag{4.4.8}$$

and

$$\tau^{right}(R_a) = \sigma^{right}(R_a) = \sigma^{left}(a) . \tag{4.4.9}$$

If the homomorphism $T : A \to B$ is bounded below then by (3.9.8) there is inclusion

$$\tau_A^{left}(a) \subseteq \tau_B^{left}(Ta) ; \ \tau_A^{right}(a) \subseteq \tau_B^{right}(Ta) . \tag{4.4.10}$$

Similarly if $T : A \to B$ is one one then by (3.9.9)

$$\pi_A^{left}(a) \subseteq \pi_B^{left}(Ta) ; \ \pi_A^{right}(a) \subseteq \pi_B^{right}(Ta) . \tag{4.4.11}$$

4.5 Boundary Spectrum

In a Banach algebra the invertible group, and more generally the semigroups of left and of right invertibles, are open sets, and it turns out that the same is true for complements of the sets of left and of right topological zero divisors: for example if $T : X \to Y$ is bounded between normed spaces there is implication

$$k\|x\| \leq \|Tx\|, \|T' - T\| < k \implies k'\|x\| \leq \|T'x\| , \tag{4.5.1}$$

with $k' = k - \|T' - T\|$. When X and Y are complete then by (3.5.10) and (3.5.11)

$$T \text{ bounded below and dense} \implies T \text{ onto},\tag{4.5.2}$$

and also, recalling (3.5.11),

$$(T \text{ bounded below}, \ T_n \text{ dense}, \ \|T_n - T\| \to 0) \implies T \text{ onto}.\tag{4.5.3}$$

This has something to say about the *topological boundary* of the invertible group and its relatives. If $a \in A$ then

$$\partial \sigma^{left}(a) \subseteq \tau^{right}(a), \ \partial \sigma^{right}(a) \subseteq \tau^{left}(a),\tag{4.5.4}$$

and hence, using (2.6.16),

$$\partial \sigma(a) \subseteq \tau^{left}(a) \cap \tau^{right}(a).\tag{4.5.5}$$

We can improve on this if we replace the topological boundary by the *fat boundary*

$$\partial(a) \equiv \partial_A(a) = \{\lambda \in \mathbb{C} : a - \lambda \in \partial A^{-1}\}.\tag{4.5.6}$$

Now

$$\partial \sigma(a) \subseteq \partial(a) \subseteq \tau^{left}(a) \cap \tau^{right}(a).\tag{4.5.7}$$

Also, with

$$\partial^{left}(a) = \{\lambda \in \mathbb{C} : a - \lambda \in \partial A_{right}^{-1}\}.\tag{4.5.8}$$

and

$$\partial^{right}(a) = \{\lambda \in \mathbb{C} : a - \lambda \in \partial A_{left}^{-1}\},\tag{4.5.9}$$

there is inclusion

$$\partial \sigma^{right}(a) \subseteq \partial^{left}(a) \subseteq \tau^{left}(a)\tag{4.5.10}$$

and

$$\partial \sigma^{left}(a) \subseteq \partial^{right}(a) \subseteq \tau^{right}(a).\tag{4.5.11}$$

If the Banach algebra homomorphism $T : A \to B$ is bounded below then this, with the inclusion (4.4.10), gives

$$\partial \sigma_A(a) \subseteq \partial_A(a) \subseteq \sigma_B^{left}(Ta) \cap \sigma_B^{right}(Ta) \subseteq \sigma_A(a).\tag{4.5.12}$$

If the Banach algebra homomorphism $T : A \to B$ is one one then there is inclusion (4.4.11), and then also

$$\text{iso } \sigma_A(a) \subseteq \text{iso}^\sim \sigma_A(a) \subseteq \sigma_B(Ta) \subseteq \sigma_A(a) , \qquad (4.5.13)$$

where (cf (2.8.4))

$$\lambda \in \text{iso}^\sim K \iff \{\lambda\} = \text{Comp}_\lambda(K) . \qquad (4.5.14)$$

We shall describe the homomorphism $T : A \to B$ of Banach algebras as *weakly Riesz* provided

$$A^{-1} + T^{-1}(0) \subseteq \text{cl}(A^{-1}) . \qquad (4.5.15)$$

We can relax this condition, and indeed extend it to ring homomorphisms, if we replace the norm closure by the spectral closure of (3.2.2). We shall say that $T : A \to B$ has the strong Riesz property provided, for arbitrary $a \in A$, there is inclusion

$$\partial \sigma_A(a) \subseteq \sigma_B(Ta) \cup \text{iso } \sigma_A(a) . \qquad (4.5.16)$$

Evidently

$$\textit{strongly Riesz} \implies \textit{Riesz} \implies \textit{weakly Riesz} ,$$

where $T : A \to B$ is (cf (4.3.12)) to be *Riesz* provided

$$A^{-1} + T^{-1}(0) \subseteq \text{QP}(A) . \qquad (4.5.17)$$

If $a = a^* \in A$ is self-adjoint in a C* algebra A then of course

$$\sigma(a) \subseteq \mathbb{R}$$

and hence

$$\sigma(a) = \partial \sigma(a) \subseteq \tau^{left}(a) \cap \tau^{right}(a) . \qquad (4.5.18)$$

It follows that there is indeed "spectral permanence": if $T : A \to B$ is an isometric *homomormorphism of C* algebras then

$$a \in A \implies \sigma_B(Ta) = \sigma_A(a) . \qquad (4.5.19)$$

4.6   Exponential Spectrum

If A is a topological semigroup with an open invertible group A^{-1} we shall write

$$A_0^{-1} = \mathrm{Comp}_1(A^{-1}) \tag{4.6.1}$$

for the connected component of the identity $1 \in A^{-1}$: we claim that

$$a \in A_0^{-1} \implies a \cdot A_0^{-1} = A_0^{-1} \cdot a \,, \tag{4.6.2}$$

so that A_0^{-1} is a normal subgroup of A^{-1}. When A is a complex Banach algebra then

$$A_0^{-1} = \mathrm{Exp}(A) = \{e^{c_1} e^{c_2} \dots e^{c_n} : n \in \mathbb{N}, c \in A^n\} \tag{4.6.3}$$

is the subgroup of A^{-1} generated by the exponentials

$$e^A = \{e^c : c \in A\} : \tag{4.6.4}$$

to see this observe that $\mathrm{Exp}(A)$ is connected and open while $A^{-1} \setminus \mathrm{Exp}(A)$ is also open in A. The group of *generalized exponentials* gives rise to the "exponential spectrum":

$$\varepsilon(a) \equiv \varepsilon_A(a) = \{\lambda \in \mathbb{C} : a - \lambda \notin \mathrm{Exp}(A)\} \,. \tag{4.6.5}$$

Evidently

$$\sigma(a) \subseteq \varepsilon(a) \subseteq \eta\sigma(a) \,. \tag{4.6.6}$$

and hence

$$\partial\varepsilon(a) \subseteq \sigma(a) \subseteq \varepsilon(a) \subseteq \eta\sigma(a) \,. \tag{4.6.7}$$

We shall write

$$\kappa(A) = A^{-1}/A_0^{-1} \tag{4.6.8}$$

for the *abstract index group* of the topological semigroup A.

When $T : A \to B$ is a continuous homomorphism of topological semigroups then as well as the inclusion (1.5.3) there is inclusion

$$T(A_0^{-1}) \subseteq B_0^{-1} \,, \tag{4.6.9}$$

and hence there is induced a homomorphism

$$\kappa(T) : \kappa(A) \to \kappa(B) \tag{4.6.10}$$

of index groups. We shall say that the homomorphism $T : A \to B$ has the *Arens property* when $\kappa(T)$ is one one, and the *Royden property* when $\kappa(T)$ is onto. Thus the Arens property takes the form

$$A^{-1} \cap T^{-1} B_0^{-1} \subseteq A_0^{-1} , \tag{4.6.11}$$

while the Royden property takes the form

$$B^{-1} \subseteq T(A^{-1}) \cdot B_0^{-1} . \tag{4.6.12}$$

All this terminology is motivated by the "Arens-Royden theorem", which says that the Gelfand homomorphism on a commutative Banach algebra has both the Arens and the Royden property. We remark that when $A = C(X)$ for a compact Hausdorff space X then the abstract index group $\kappa(A)$ can be identified with the "first Cech cohomology group" of the space X.

Of course if a bounded homomorphism $T : A \to B$ is onto then

$$T \operatorname{Exp}(A) = \operatorname{Exp}(B) . \tag{4.6.13}$$

It is *Kuiper's theorem* that if $A = B(X)$ for a Hilbert space X then the invertible group A^{-1} is contractible, in the sense of §2.7, so that of course

$$B(X)^{-1} = \operatorname{Exp} B(X) . \tag{4.6.14}$$

Equality (4.6.14) continues to hold when $X = \ell_p$, with $1 \le p \le \infty$, and with $X = c_0$. We shall find, however, looking at block matrices, that it fails when for example $X = \ell_p \oplus \ell_q$ with $p \ne q$. This because that when $p \ne q$ the spaces $Y = \ell_p$ and $Z = \ell_q$ are mutually "incomparable", in the sense that everything in either $BL(Y, Z)$ or $BL(Z, Y)$ is "inessential". We have

$$B(X) = \begin{pmatrix} B(Y) & BL(Z, Y) \\ BL(Y, Z) & B(Z) \end{pmatrix} , \tag{4.6.15}$$

and hence, taking Calkin quotients,

$$A = [B(Y)] , \, M = [BL(Z, Y)] , \, N = [BL(Y, Z)] , \, B = [B(Z)] , \tag{4.6.16}$$

we have

$$I - MN \subseteq A^{-1} \text{ and } I - NM \subseteq B^{-1} . \tag{4.6.17}$$

It follows that every $[T] \in G = [B(X)]$ is a "spectral diagonal" in the sense (1.11.16), so that by (1.11.10)

$$[T] \in G^{-1} \Longleftrightarrow (a \in A^{-1} \text{ and } b \in B^{-1}) . \qquad (4.6.18)$$

It follows that also

$$\mathrm{Exp}(G) = \begin{pmatrix} \mathrm{Exp}(A) & M \\ N & \mathrm{Exp}(B) \end{pmatrix} . \qquad (4.6.19)$$

For a specific example take

$$T = U = \begin{pmatrix} u & w \\ 0 & v \end{pmatrix} , \; T^* = V = U^* = \begin{pmatrix} u^* & 0 \\ w^* & v^* \end{pmatrix} , \qquad (4.6.20)$$

where $u : Y \to Y$ and $v : Z \to Z$ are the forward shifts on ℓ_p and ℓ_q while $w : Z \to Y$ and $w^* : Y \to Z$ are the formally same, rank one projection: thus

$$u^*u = 1 = uu^* + ww^* \in B(Y) , \; vv^* = 1 = v^*v + w^*w \in B(Z) , \qquad (4.6.21)$$

giving

$$U^*U = I = UU^* . \qquad (4.6.22)$$

We claim

$$U \in B(X)^{-1} \setminus \mathrm{Exp}\, B(X) . \qquad (4.6.23)$$

Certainly (4.6.22) says that U is invertible; on the other hand, by (4.6.13), there is implication

$$U \in \mathrm{Exp}\, B(X) \Longrightarrow [U] \in \mathrm{Exp}(G) , \qquad (4.6.24)$$

and then, with $a = [u]$ and $b = [v]$,

$$[T] = [U] \in \mathrm{Exp}(G) \Longleftrightarrow (a \in \mathrm{Exp}(A) \text{ and } b \in \mathrm{Exp}(B)) . \qquad (4.6.25)$$

But

$$a \in \mathrm{Exp}(A) \Longrightarrow \mathrm{index}(u) = 0 \, ; \; b \in \mathrm{Exp}(B) \Longrightarrow \mathrm{index}(v) = 0, \qquad (4.6.26)$$

which is obviously false. Alternatively, if a and b were generalized exponentials then u and v would have invertible generalized inverses as well as one-sided inverses, and hence (1.2.16) themselves be invertible.

4.7 Essential Spectrum

A bounded linear operator $T \in BL(X, Y)$ between Banach spaces is said to be *Fredholm* when

$$\dim T^{-1}(0) < \infty , \tag{4.7.1}$$

$$\dim(Y/\text{cl } TX) < \infty \tag{4.7.2}$$

and also

$$T(X) = \text{cl } TX . \tag{4.7.3}$$

We remark that, with the aid of the open mapping theorem, that the condition

$$\dim(Y/TX) < \infty \tag{4.7.4}$$

implies the closed range condition (4.7.3). By *Atkinson's theorem* the condition

$$T + BL_0(X, Y) \in (BL(X, Y)/BL_0(X, Y))^{-1} \tag{4.7.5}$$

implies that T is Fredholm in the sense (4.7.1)–(4.7.3), which in turn (1.9.3) implies

$$T + BL_{00}(X, Y) \in (BL(X, Y)/BL_{00}(X, Y))^{-1} . \tag{4.7.6}$$

Thus the "spatially Fredholm" conditions (4.7.1)–(4.7.3) coincide with the condition (1.5.6) for each of two *Calkin homomorphisms*; (1.9.5) holds with $J = B_{00}(X)$ and $K = B_0(X)$.

When $T \in BL(X, Y)$ is Fredholm we shall write

$$\text{index}(T) = \dim T^{-1}(0) - \dim(Y/(\text{cl } TX)) ; \tag{4.7.7}$$

now it is *Schechter's theorem* that

$$\text{index}(T) = 0 \tag{4.7.8}$$

is equivalent to the Weyl condition (1.9.1) for each of the two Calkin homomorphisms. The *logarithmic law* of the index says that (1.9.8) holds: if $T \in BL(X, Y)$ and $S \in BL(Y, Z)$ are both Fredholm then the index of the product ST is the sum of the indexes of the factors S and T. We remark that the closed range condition (4.7.3) is important here: if we define $T \in BL(X, Y)$ to be weakly Fredholm if it satisfies the conditions (4.7.1) and (4.7.2), but possibly not (4.7.3), and define the index by the formula (4.7.7), then the logarithmic law (1.9.8) may possibly fail.

When $Y = X$ then the condition that $T \in BL(X, Y) = B(X)$ is Fredholm with

$$\max(\operatorname{ascent}(T), \operatorname{descent}(T)) < \infty \tag{4.7.9}$$

is equivalent to the Browder condition (1.9.2) for each of the two Calkin homomorphisms.

When $T \in B(X)$ we shall write

$$\sigma_{ess}(T) = \{\lambda \in \mathbb{C} : T - \lambda I \text{ not Fredholm}\} \tag{4.7.10}$$

for the (Fredholm) *essential spectrum* of T: this is of course also the spectrum $\sigma_A(a)$ of the Calkin quotient

$$a = T + B_0(X) \in A = B(X)/B_0(X) .$$

It is therefore clear that the essential spectrum obeys the spectral mapping theorem for polynomials. The *Weyl essential spectrum* is

$$\omega_{ess}(T) = \{\lambda \in \mathbb{C} : T - \lambda I \text{ not Fredholm of index zero}\} . \tag{4.7.11}$$

For this we get only half the spectral mapping theorem:

$$\omega_{ess} p(T) \subseteq p\omega_{ess}(T) . \tag{4.7.12}$$

The *Browder essential spectrum* is

$$\beta_{ess}(T) = \{\lambda \in \mathbb{C} : T - \lambda I \text{ not Fredholm of finite ascent and descent}\} . \tag{4.7.13}$$

Now again there is equality in the spectral mapping theorem:

$$p\beta_{ess}(T) = \beta_{ess} p(T) . \tag{4.7.14}$$

The *punctured neighbourhood theorem* says that the Calkin homomorphism has the strong Riesz property (4.5.16). The argument says that if $T = TT'T \in B(X)$ is Fredholm, with a generalized inverse $T' \in B(X)$ leaving the hyperrange invariant

$$T'T^\infty(X) \subseteq T^\infty(X) ,$$

then, writing $R^\sim$ for the restriction of R to $T^\infty(X)$, if

$$(I + T'S)^\sim \in B(T^\infty(X))^{-1}$$

is invertible then there is equality

$$\dim(T - S)^{-1}(0) = \dim(T - S)^{\sim -1}(0) = \operatorname{index}(T - S)^\sim . \tag{4.7.15}$$

In particular the "nullity" $\dim(T-S)^{-1}(0)$ is independent of S for sufficiently small commuting invertible $S \in B(X)$. Specializing to scalars $S = \lambda I$ now gives (4.5.16):

$$\partial \sigma(T) \subseteq \text{iso } \sigma(T) \cup \sigma_{ess}(T) \,. \tag{4.7.16}$$

The strong Riesz property gives an alternative expression for the Browder spectrum:

$$\beta_{ess}(T) = \sigma_{ess}(T) \cup \text{acc } \sigma(T) \,. \tag{4.7.17}$$

The logarithmic law of the index (1.9.8) does not survive without the closed range property, and fails for "weakly Fredholm" operators. For example if $T = W : X \to X$ is one-one and dense, but not onto, and $S = I - f \odot e$ with

$$f(Te) = 1 \text{ and } e \notin T(X) \,, \tag{4.7.18}$$

then for arbitrary $x \in X$,

$$STx = Tx - f(Tx)e \,; \; = 0 \Longrightarrow f(Tx) = 0 \Longrightarrow Tx = 0 \,. \tag{4.7.19}$$

Thus

$$(ST)^{-1}(0) = T^{-1}(0) \text{ and cl } ST(X) = \text{cl } S(X) \,, \tag{4.7.20}$$

and hence

$$\text{index}(ST) = 0 - 1 \neq 0 + 0 = \text{index}(S) + \text{index}(T) \,. \tag{4.7.21}$$

4.8 Local Spectrum

If $f : X \to X$ is a mapping of sets then we can *iterate*, setting

$$f^0 = I : X \to X \tag{4.8.1}$$

to be the identity mapping $x \mapsto x$ and then, for each $n \in \infty = \{0, 1, 2, 3, \ldots\}$,

$$f^{n+1} = f \circ f^n \,. \tag{4.8.2}$$

Evidently this formula would make sense for more general *ordinals* n; it would be less clear how to interpret for example f^∞ for *limit ordinals* such as ∞. We can however interpret *transfinite ranges*

$$f^\alpha(X) = R_\alpha(f) \,,$$

setting

$$R_\beta(f) = \bigcap_{\alpha < \beta} R_\alpha(f) \tag{4.8.3}$$

for limit ordinals β, and relying on (4.8.2) for successor ordinals. This remains valid for linear mappings $f = T$ on vector spaces X; in particular we recall the hyperrange $T^\infty(X)$ of (1.14.3), the intersection of the ranges $T^n(X)$ corresponding to $n \in \infty$: this contains a subspace known as the *coeur algébrique*. One way to characterize this is to collect all points $\xi_0 \in X$ for which there exists a sequence $\xi \in X^\infty$ with the property that, for each $n \in \infty = \{0, 1, 2, \ldots\}$

$$\xi_n = T(\xi_{n+1}) . \tag{4.8.4}$$

It turns out that this can be represented as $T^\alpha(X)$, where α is a sufficiently large *ordinal number*; alternatively the couer algébrique is the largest subspace $Y \subseteq X$ for which there is equality $T(Y) = Y \subseteq X$. The *couer analytique* or *transfinite range* is

$$T^\omega(X) = \{x \in X : \exists\, \xi \in k_\infty(X) : \xi_n = T(\xi_{n+1}) : x = \xi_0\} \tag{4.8.5}$$

where

$$k_\infty(X) = \{\xi \in X^\infty : \limsup_n \|\xi_n\|^{1/n} < \infty\} . \tag{4.8.6}$$

This coincides with the *holomorphic range* of T, which consists of those points $x \in X$ for which there exists a holomorphic function $f : U \to X$, defined on $U \in \mathrm{Nbd}(0)$, for which

$$(T - zI)f(z) \equiv x \text{ on } U . \tag{4.8.7}$$

The connection between these two characterizations is given by the expansion

$$f(z) \equiv \sum_{n=0}^{\infty} z^n \xi_n . \tag{4.8.8}$$

The *holomorphic kernel points* of T constitute the intersection

$$T^{-1}(0) \cap T^\omega(X) : \tag{4.8.9}$$

equivalently these are the points $x = g(0) \in X$ for which there is holomorphic $g : U \to X$ with $U \in \mathrm{Nbd}(0)$ for which

$$(T - zI)g(z) \equiv 0 \text{ on } U . \tag{4.8.10}$$

When

$$T^{-1}(0) \cap T^{\omega}(X) = \{0\} \tag{4.8.11}$$

we shall say that T is *locally one one*, or alternatively that T "has the single-valued extension property at 0". The *local eigenvalues* of $T \in B(X)$ are now given by

$$\pi_{loc}(T) = \{\lambda \in \mathbb{C} : (T - \lambda I)^{-1}(0) \cap (T - \lambda I)^{\omega}(X) \neq \{0\}\}. \tag{4.8.12}$$

We recall that $T \in A = L(X)$ is said to be *Kato invertible* if it is relatively regular and (splitting) hyperexact in the sense of the last paragraph of Sect. 1.10; when $A = B(X)$ for a Banach space X then of course we require that the generalized inverse of $T \in A$ be also in $B(X)$. For $T \in B(X)$ there is implication

$$\textit{consortedly regular} \implies \textit{hyperregular} \implies \textit{holomorphically regular}. \tag{4.8.13}$$

Here $a \in A$ is *consortedly regular* if there are sequences (c_n) in $\text{comm}^{-1}(a)$ and (a'_n) in $A^{\cap}$ for which

$$\|c_n\| + \|a'_n - a'\| \to 0 \text{ and } a - c_n = (a - c_n)a'_n(a - c_n), \tag{4.8.14}$$

and is *holomorphically regular* if there is $U \in \text{Nbd}(0)$ and a holomorphic mapping $a'_z : U \to A$ for which

$$a - z \equiv (a - z)a'_z(a - z) : U \to A. \tag{4.8.15}$$

Now the *Kato spectrum* of $T \in B(X)$ is to be given by

$$\sigma_{Kato}(T) = \{\lambda \in \mathbb{C} : T - \lambda I \text{ not Kato invertible}\}. \tag{4.8.16}$$

More generally we shall say that $T \in B(X)$ is *Kato non singular* if it is hyperregular with closed range, and set

$$\tau_{Kato}(T) = \{\lambda \in \mathbb{C} : T - \lambda I \text{ not Kato non singular}\}. \tag{4.8.17}$$

It is easily checked that there is implication

$$T(X) = X \implies T^{\omega}(X) = T(X) = X, \tag{4.8.18}$$

and hence there is equality

$$\pi_{loc}(T) \cup \tau^{right}(T) = \sigma(T). \tag{4.8.19}$$

With the help of the enlargement, we can also introduce "local approximate eigenvalues", setting

$$\tau_{loc}(T) = \pi_{loc}\mathbf{Q}(T) .\tag{4.8.20}$$

For Banach algebra elements $a \in A$ we shall define

$$\pi_{Aloc}^{left}(a) = \pi_{loc}(L_a) ;\ \pi_{Aloc}^{right}(a) = \pi_{loc}(R_a) .\tag{4.8.21}$$

There is also another kind of "local spectrum", for operators $T \in B(X)$, induced by vectors $x \in X$: the (somewhat anomalous) notation is that

$$\lambda \notin \sigma_T(x) \Longleftrightarrow x \in (T - \lambda I)^{\omega}(X) .\tag{4.8.22}$$

4.9   Spectral Pictures

A "spectrum" is in the first instance a nonempty compact set $K \subseteq \mathbb{C}$ in the complex plane. By a *spectral landscape* we shall understand an ordered pair

$$(K, \nu)$$

in which K is a spectrum and

$$\nu : \mathrm{Hole}(K) \rightarrow \kappa(A)$$

is a mapping from (cf (2.6.9)) the holes in K into the abstract index group (4.6.8). Now the "spectral landscape" of a Banach algebra element $a \in A$ will be given by

$$(K, \nu) = (\sigma(a), \iota_\sigma(a)) ,\tag{4.9.1}$$

where

$$\iota_\sigma(a)(H) = A_0^{-1}(a - \lambda) \in \kappa(A)\ (\lambda \in H \in \mathrm{Hole}\ \sigma(a)) .\tag{4.9.2}$$

Generally if (K, ν) is a spectral landscape and $f : \mathbb{C} \rightarrow \mathbb{C}$ is a polynomial, or more generally holomorphic on a neighbourhood of ηK, we shall define the image $f(K, \nu)$ to be

$$f(K, \nu) = (f(K), \nu_f) ,\tag{4.9.3}$$

where

$$\nu_f(L) = \prod\{\nu(H)^{N_f(L,H)} : L \in \mathrm{Hole}\ f(K), L \subseteq f(H)\} ,\tag{4.9.4}$$

where

$$N_f(L, H) = \#\{\lambda \in H : f(\lambda) = \mu\} = \#f^{-1}(\mu) \cap H \ . \tag{4.9.5}$$

The point of all this is that there is a spectral mapping theorem for pictures:

$$(\sigma f(a), \iota_\sigma f(a)) = (f\sigma(a), \iota_\sigma(a)_f) \ . \tag{4.9.6}$$

The fundamental example is the Calkin algebra $A = B(X)/B_0(X)$ got by quotienting out the compact operators on a Banach space, where there is a natural mapping

$$\text{Index} : \kappa(A) \to \mathbb{Z} \tag{4.9.7}$$

derived from the Fredholm index.

We remark that the abstract index group $\kappa(A)$ can be presented as the intersection of a left and a right index semigroup, with

$$\kappa_{left}(A) = A_{left}^{-1}/A_0^{-1} = \{A_0^{-1}a : a \in A_{left}^{-1}\} \tag{4.9.8}$$

and

$$\kappa_{right}(A) = A_{right}^{-1}/A_0^{-1} = \{aA_0^{-1} : a \in A_{right}^{-1}\} \ . \tag{4.9.9}$$

4.10 Müller Regularity

There is a correspondence

$$\omega \longleftrightarrow H_\omega \tag{4.10.1}$$

between a "spectrum" $a \mapsto \omega(a) \subseteq \mathbb{C}$ and a "regularity" $H_\omega \subseteq A$ given by setting

$$\omega(a) = \{\lambda \in \mathbb{C} : a - \lambda \notin H_\omega\} \ ; \tag{4.10.2}$$

equivalently

$$H_\omega = \{a \in A : 0 \notin \omega(a)\} \ . \tag{4.10.3}$$

For example

$$\omega = \sigma \implies H_\omega = A^{-1} \ .$$

For a mapping $a \mapsto \omega(a)$ to behave like a spectrum the subset $H_\omega = R$ should be a *regularity* in the following sense:

$$R \neq \emptyset ; \tag{4.10.4}$$

for arbitrary $a \in A$ and $n \in \mathbb{N}$

$$a \in R \Longleftrightarrow a^n \in R ; \tag{4.10.5}$$

for arbitrary commutative $\{a, a', b, b'\} \subseteq A$

$$\left(b'b + aa' = 1\right) \Longrightarrow \left(ba \in R \Longleftrightarrow \{a, b\} \subseteq R\right) . \tag{4.10.6}$$

Partially more generally, (1.10.12) says that the relatively regulars $H = A^\cap$ define what we shall call a "non commutative regularity"; the same is true of A_{left}^{-1}, A_{right}^{-1} and the two-sided invertibles A^{-1}. For example the argument for left invertibles is a beefed-up version of (1.2.10). The monomorphisms A_{left}^o and the epimorphisms A_{right}^o of (1.3.2) are also regularities. In a normed algebra or more generally a normed linear category A, the strong monomorphisms $A_{left}^\bullet$ and $A_{right}^\bullet$ of (3.9.1) and (3.9.2) are regularities. Evidently if R is a regularity then $A^{-1} \subseteq R$, and hence if H_ω is a regularity then $\omega(a) \subseteq \sigma(a)$. The *spectral mapping theorem* holds for a spectrum derived from a regularity:

$$f \in \mathrm{Holo}_1 \sigma(a) \Longrightarrow \omega f(a) = f\omega(a) . \tag{4.10.7}$$

Here

$$f \in \mathrm{Holo}_1(K) \tag{4.10.8}$$

means that $f : U \to \mathbb{C}$ is defined and holomorphic on, and non constant on each connected component of, an open neighbourhood $U \in \mathrm{Nbd}(K)$. When $f = p$ is a nonconstant polynomial and $\mu \in \mathbb{C}$ we argue that there are $k \in \mathbb{N}$ and a polynomial q with $q(\lambda) \neq 0$ for which

$$\mu = p(\lambda) \Longrightarrow p(z) - \mu \equiv (z - \lambda)^k q(z) , \tag{4.10.9}$$

and hence, by the *Euclidean algorithm*, polynomials q' and r for which

$$(z - \lambda)^k r(z) - q'(z)q(z) \equiv 1 , \tag{4.10.10}$$

and therefore also

$$(a - \lambda)^k r(a) - q'(a)q(a) = 1 \in A . \tag{4.10.11}$$

Generally we can write $f = pg$ with a polynomial p and a holomorphic function $g \in \text{Holo } \sigma(a)$ for which

$$\sigma(a) \cap g^{-1}(0) = \emptyset . \tag{4.10.12}$$

More generally we shall describe $\omega : a \mapsto \omega(a)$ as a *Möbius spectrum* if the right hand side of (4.10.7) holds whenever

$$f \equiv \frac{\alpha z + \beta}{\gamma z + \delta} \tag{4.10.13}$$

is well defined on $\sigma(a) \cup \omega(a)$. For example (2.6.25) and (2.6.26)

$$\partial \sigma f(a) \subseteq f \partial \sigma(a) \tag{4.10.14}$$

while

$$f \eta \sigma(a) \subseteq \eta \sigma f(a) . \tag{4.10.15}$$

There is a more primitive idea of "regularity": in an additive category A we shall call $H \subseteq A$ a *non commutative Müller regularity* if there is implication, for arbitrary splitting exact $(b, a) \in A^2$,

$$ba \in H \Longleftrightarrow \{a, b\} \subseteq H . \tag{4.10.16}$$

Since relative regularity is, by (1.10.12), a non commutative regularity, it follows that Kato invertibility is a Müller regularity. It is also true that Kato non singularity is a Müller regularity; this is because, by (3.5.18) and (3.5.19), the closed range condition is another non commutative regularity. The locally one-one property of (4.8.11) is also a Müller regularity; the story begins with the ascent and descent one property of (1.14.7) and (1.14.8). and the consequences of middle non singularity (1.14.13), which show that finite ascent and finite descent are Müller regularities. To extend this to local one-one-ness we have to replace the ranges in (1.14.18) and (1.14.19) by couer analytique.

4.11 Numerical Range

A *hermitian subspace* H of a complex linear algebra A is a real-linear subspace $H \subseteq A$ for which

$$H \cap i H = O \equiv \{0\} \tag{4.11.1}$$

with

$$1 \in H . \tag{4.11.2}$$

For example the real scalars $H = \mathbb{R} \subseteq A$ constitute a hermitian subspace; the intersection of two hermitian subspaces is hermitian, and an easy Zorn lemma argument shows that every hermitian subspace is contained a maximal hermitian subspace. Hermitian subspaces give rise to complex Palmer subspaces $H + iH \subseteq A$ carrying *involutions* almost in the sense of (1.5.21): we define $* : H + iH \to H + iH$ by setting

$$(h + ik)^* = h - ik \ (h, k \in H) \,. \tag{4.11.3}$$

Evidently, for arbitrary $x, y \in H + iH$ and $\alpha, \beta \in \mathbb{C}$

$$(\alpha x + \beta y)^* = \overline{\alpha} x^* + \overline{\beta} y^* \,; \ (x^*)^* = x \,; \ 1^* = 1 \,. \tag{4.11.4}$$

Conversely a partially defined involution $* : K \to K \subseteq A$ on a complex-linear subspace gives rise to a hermitian subspace $H = \{a \in K : a^* = a\}$. When A is a topological algebra then it is desirable that $H + iH$ is norm closed, and also that $*$ is continuous: in a Banach algebra this asks for a certain mutual "orthogonality" between H and iH.

It is not clear that the Palmer subspace is closed under multiplication. Generally, if $a, b \in H + iH$ then

$$\{ab, b^* a^*\} \subseteq H + iH \iff \{ab + b^* a^*, i(ab - b^* a^*)\} \subseteq H \,. \tag{4.11.5}$$

We shall call the hermitian subspace $H \subseteq A$ a *Lie subspace* if

$$i[H, H] \equiv \{i(ab - ba) : a, b \in H\} \subseteq H \,, \tag{4.11.6}$$

and a *Jordan subspace* if instead

$$]H, H[\equiv \{ab + ba : a, b \in H\} \subseteq H \,. \tag{4.11.7}$$

Necessary and sufficient for the Palmer subspace $H + iH$ to be a subalgebra is that H is both a Lie and a Jordan subspace, in which case

$$(ab)^* = b^* a^* \ (a, b \in H + iH) \,. \tag{4.11.8}$$

In a Banach algebra A the *Vidav hermitian* elements are determined by the *numerical range*: for $a \in A$

$$V_A(a) = \{\varphi(a) : \varphi \in \text{State}(A)\} \tag{4.11.9}$$

is defined, with of course A^* the Banach space dual of A, by the "states"

$$\text{State}(A) = \{\varphi \in A^* : \|\varphi\| = 1 = \varphi(1)\} \,. \tag{4.11.10}$$

Specifically, elements with real numerical range are said to be *hermitian*:

$$\text{Re}(A) = \{a \in A : V_A(a) \subseteq \mathbb{R}\} ; \tag{4.11.11}$$

equivalently

$$\lim_{t \to 0} \frac{\|1 + ita\| - 1}{t} = 0 ; \tag{4.11.12}$$

equivalently

$$\forall t \in \mathbb{R} : \|e^{ita}\| = 1 . \tag{4.11.13}$$

To properly understand Vidav hermitians we need both the first and the third of these conditions. The same argument that shows their equivalence also shows that $\text{Re}(A)$ is a Lie subspace, and hence that the Palmer subspace

$$\text{Reim}(A) = \text{Re}(A) + i \, \text{Re}(A) \tag{4.11.14}$$

is a Lie subalgebra of A. The norm-closedness of $\text{Reim}(A)$ is fall out from the "Phragmen-Lindelof theorem" and maximum modulus: if $(h, k) \in \text{Re}(A)^2$ then

$$\|(h + ik)^*\| \leq \|h\| + \|k\| \leq 2e\|h + ik\| .$$

It is the *Vidav-Palmer theorem* that necessary and sufficient for A to be a C* algebra is that

$$A = \text{Reim}(A) . \tag{4.11.15}$$

If in particular A is a C* algebra then, in the notation of (4.3.26),

$$A^+ = \{a \in \text{Re}(A) : \sigma_A(a) \subseteq \mathbb{R}^+ = [0, \infty)\} . \tag{4.11.16}$$

The extreme points of the convex set $\text{State}(A)$ are known as the *Pure states*.

4.12 Invariant Subspaces

For bounded linear operators $T \in B(X)$ on Banach spaces we insist that "invariant subspaces" $T(Y) \subseteq Y \subseteq X$ be norm *closed*. From (1.12.13)–(1.12.16) it follows that the three spectrums

$$\sigma(T) ; \; \sigma(T_Y) ; \; \sigma(T_{/Y}) , \tag{4.12.1}$$

form a "love knot": each is included in the union of the other two. We shall call the invariant subspace $Y \subseteq X$ *spectrally invariant* for $T \in B(X)$ whenever

$$\sigma(T_Y) \cap \sigma(T_{/Y}) = \emptyset . \tag{4.12.2}$$

A spectrally invariant subspace Y will always be *reducing*, in the sense that it will also be complemented, with an invariant complement, $Z \subseteq X$ for which:

$$X = Y + Z ; \; Y_\cap Z = \{0\} ; \; T(Z) \subseteq Z ; \tag{4.12.3}$$

equivalently there will be $P \in B(X)$ for which

$$P^2 = P ; \; PT = TP . \tag{4.12.4}$$

Spectrally invariant subspaces $Y \subseteq X$ are also *hyperinvariant*, in the sense that there is implication

$$S \in \text{comm}(T) \Longrightarrow S(Y) \subseteq Y . \tag{4.12.5}$$

More generally we shall say that $Y \subseteq X$ is "comm^2 invariant" provided there is implication

$$S \in \text{comm}^2(T) \Longrightarrow S(Y) \subseteq Y ; \tag{4.12.6}$$

more generally still we shall say that $Y \subseteq X$ is *inverse invariant* if there is implication, for $\lambda \in \mathbb{C}$,

$$S = T - \lambda I \in B(X)^{-1} \Longrightarrow S^{-1}(Y) \subseteq Y . \tag{4.12.7}$$

Obviously

$$hyperinvariant \Longrightarrow comm^2 \; invariant \Longrightarrow inverse \; invariant \Longrightarrow invariant .$$

It turns out that none of these implications are reversible; nor do hyperinvariant and reducing together imply spectrally invariant.

If we write, for linear $T : X \to X$,

$$\pi(T) = \{\lambda \in \mathbb{C} : (T - \lambda I)^{-1}(0) \neq \{0\}\} \tag{4.12.8}$$

for the "spatial eigenvalues", and

$$\pi'(T) = \{\lambda \in \mathbb{C} : (T - \lambda I)(X) \neq X\} , \tag{4.12.9}$$

then evidently

$$\sigma(T) = \pi(T) \cup \pi'(T) \qquad (4.12.10)$$

The *algebraic vectors* for $T : X \to X$ make up the hyper invariant subspace

$$E_X(T) = \sum_{\lambda \in \mathbb{C}} (T - \lambda I)^{-\infty}(0) ; \qquad (4.12.11)$$

equivalently

$$E_X(T) = \bigcup \{p(T)^{-1}(0) : 0 \neq p \in \text{Poly}\} . \qquad (4.12.12)$$

When

$$E_X(T) = X \qquad (4.12.13)$$

we say that T is *locally algebraic*. When $T \in B(X)$ is bounded on a Banach space then it is *Kaplansky's lemma* that

$$T \text{ locally algebraic} \implies T \text{ algebraic} ,$$

in the sense that there is $0 \neq p \in \text{Poly}$ for which

$$p(T) = 0 \in B(X) . \qquad (4.12.14)$$

Lomonosov's Lemma tells us that non scalar compact operators on Banach spaces have (non trivial closed) hyperinvariant subspaces: if $A \subseteq B(X)$ is a subalgebra for which there is implication, for closed subspaces $Y \subseteq X$,

$$AY \subseteq Y \implies Y \in \{O, X\} , \qquad (4.12.15)$$

then there is also implication

$$0 \neq K \in B_0(X) \implies \exists T \in A : (I + TK)^{-1}(0) \neq \{0\} . \qquad (4.12.16)$$

In a curious converse, it has been shown that if the algebra $A \subseteq B(X)$ does have an invariant subspace then, whenever

$$T \in A \text{ with } 0 \notin \text{cvx } \sigma(T) ,$$

then there exists $S \in B(X)$ for which

$$\dim(ST + TS)(X) = 1 \text{ and } \forall R \in A : I + RS \in B(X)^{-1} . \qquad (4.12.17)$$

Complemented or not, invariant subspaces also induce block structure: if $T(Y) \subseteq Y \subseteq X$, we have the family

$$T_U = \begin{pmatrix} T_Y & U \\ 0 & T_{/Y} \end{pmatrix} : \begin{pmatrix} Y \\ X/Y \end{pmatrix} \rightarrow \begin{pmatrix} Y \\ X/Y \end{pmatrix} \tag{4.12.18}$$

indexed by

$$U \in BL(X/Y, Y) ; \tag{4.12.19}$$

in the bottom left hand corner

$$K_Y T J_Y = T_{/Y} K_Y J_Y = K_Y J_Y T_Y = 0 \in BL(Y, X/Y) . \tag{4.12.20}$$

If $f \in \mathrm{Holo}(\sigma(T_Y) \cup \sigma(T_{/Y}))$ then, with

$$T'_U = \begin{pmatrix} T_Y & T_Y U - U T_Y \\ 0 & T_{/Y} \end{pmatrix} , \quad Q_Y = \begin{pmatrix} I_Y & U \\ 0 & 0_{/Y} \end{pmatrix} , \tag{4.12.21}$$

we have

$$f(T_U) = \begin{pmatrix} f(T_Y) & f(T_Y)U - Uf(T_Y) \\ 0 & f(T_{/Y}) \end{pmatrix} , \tag{4.12.22}$$

Spectral disjointness (4.12.2) is necessary and sufficient for

$$Q_U \in \mathrm{Holo}(T'_U) , \tag{4.12.23}$$

while weaker left, right disjointness conditions give

$$Q_U \in \mathrm{comm}^2(T'_U) . \tag{4.12.24}$$

4.13   Peripheral Spectrum

Generally the *peripheral spectrum* of $a \in A$ is given by

$$\sigma_{per}(a) = \{\lambda \in \sigma(a) : |\lambda| = |a|_\sigma\} . \tag{4.13.1}$$

This is particularly interesting when the algebra A is "partially ordered".

A *partially ordered* ring A, with identity 1, has distinguished a *positive cone* $A^+ \subseteq A$ with the properties

$$1 \in A^+ + A^+ \subseteq A^+ ; \tag{4.13.2}$$

$$A^+ \cap (-A^+) = \{0\} ; \tag{4.13.3}$$

$$A^+ \cdot A^+ \subseteq A^+ . \tag{4.13.4}$$

The "order" is imposed by the declaration, for $\{a, b\} \subseteq A$, that

$$a \leq b \iff b - a \in A^+ .$$

When A is a real or a complex algebra we also ask

$$\mathbb{K}^+ A^+ \subseteq A^+ . \tag{4.13.5}$$

Slightly more generally, we can relax the multiplicative property (4.13.4) (*cf* (1.4.5), (1.9.2)): in a *commutatively ordered* ring A,

$$A^+ \cdot_{comm} A^+ \subseteq A^+ . \tag{4.13.6}$$

When A is a topological ring, in particular a Banach algebra, we also ask that the positive cone be norm-closed:

$$cl\ A^+ \subseteq A^+ . \tag{4.13.7}$$

We remark that the positive cone (4.3.24) of a C* algebra in general fails to satisfy the multiplicative property (4.13.4), but does obey the commutative version (4.13.6): we only ask that commuting products of positives are positive. The failure for the modulus (3.11.19)

$$a \mapsto |a| = (a^*a)^{1/2}$$

of the triangle inequality makes the additive property (4.13.2) less than obvious from (4.3.26); it becomes much clearer with (4.11.15).

The *Perron-Frobenius theorem* says that in a (commutatively) partially ordered Banach algebra A the spectral radius of an element $a \in A$ is an eigenvalue:

$$|a|_\sigma \in \pi^{left}(a) . \tag{4.13.8}$$

We can also show, using the Schauder fixed point theorem, that there is a positive eigenvector here:

$$b = a - |a|_\sigma \implies L_b^{-1}(0) \cap A^+ \neq \{0\} . \tag{4.13.9}$$

Several Variables

5

If the spectrum of a linear operator or algebra element is a subset of the complex numbers, then we may expect the spectrum of a pair of operators be a set of pairs of complex numbers.

5.1 Non Commutative Polynomials

If the "spectrum" of a single element $a \in A$ is derived from invertibility, then the "joint spectrum" of a tuple $a = (a_1, a_2, \ldots, a_n) \in A^n$ is derived from some kind of "joint invertibility": if A is a ring we shall write

$$A_{left}^{-n} = \{a \in A^n : 1 \in \sum_{j=1}^{n} A a_j\} \tag{5.1.1}$$

and

$$A_{right}^{-n} = \{a \in A^n : 1 \in \sum_{j=1}^{n} a_j A\} \tag{5.1.2}$$

for the "jointly left" and "jointly right" invertible n tuples in A^n, so that $a \in A^n$ does not generate a proper left, or right, ideal in A. Equivalently there are $a' \in A^n$ and $a'' \in A^n$ for which

$$a' \cdot a \equiv \sum_{j=1}^{n} a_j' a_j = 1 \text{ and } \sum_{j=1}^{n} a_j a_j'' \equiv a \cdot a'' = 1 . \tag{5.1.3}$$

When A is a complex linear algebra this gives rise to left and right spectrum, setting

$$\sigma_A^{left}(a) = \{\lambda \in \mathbb{C}^n : a - \lambda \notin A_{left}^{-n}\} \tag{5.1.4}$$

and

$$\sigma_A^{right}(a) = \{\lambda \in \mathbb{C}^n : a - \lambda \notin A_{right}^{-n}\} . \tag{5.1.5}$$

If $z = (z_1, z_2, \ldots, z_n) : \mathbb{C}^n \to \mathbb{C}^n$ are the coordinates on the space $\mathbb{C}^n$ then $\mathbb{C}[z]$ will be "polynomials in n complex variables"; then instead by Poly_n we shall understand *non commutative* polynomials in n variables, that is to say the free algebra on the n generators $z = (z_1, z_2, \ldots, z_n)$; more generally Poly_n^m will represent m-tuples of such non commutative polynomials . Now $p \in \text{Poly}_n^m$ induces a well defined mapping $a \mapsto p(a)$ from A^n to A^m for a complex linear algebra A, and there is a one way spectral mapping theorem: for $p \in \text{Poly}_n^m$ and $a \in A^n$

$$p\omega(a) \subseteq \omega p(a) \subseteq \mathbb{C}^m , \tag{5.1.6}$$

whenever

$$\omega \in \{\sigma^{left}, \sigma^{right}\} . \tag{5.1.7}$$

(5.1.6) follows from the *remainder theorem* for noncommutative polynomials, which says that if $p \in \text{Poly}_n$ and $\lambda \in \mathbb{C}^n$ we can write, for $p \in \text{Poly}_n$

$$p(z) - p(\lambda) \equiv (z - \lambda) \cdot q_\lambda(z) \equiv r_\lambda(z) \cdot (z - \lambda) \tag{5.1.8}$$

with polynomials q_λ, r_λ in Poly_n^n. To establish (5.1.8) we observe it is clear if either $p = p(0)$ is a constant or $p = z_j$ is a coordinate, and then note that if (5.1.8) holds for each of two polynomials $p = p'$ and $p = p''$ then it continues to hold for the sum polynomial $p = p' + p''$ and the product polynomial $p = p'p''$.

In general in neither case of (5.1.7) can the inclusion (5.1.6) can be replaced by equality, even in the algebra of 2×2 complex matrices: for if we set (generating the *special Lie algebra*),

$$a = (a_1, a_2) \text{ with } a_1 = \begin{pmatrix} 0 & 1 \\ 0 & 0 \end{pmatrix} \quad a_2 = \begin{pmatrix} 0 & 0 \\ 1 & 0 \end{pmatrix}, \tag{5.1.9}$$

then by (5.1.6) there is inclusion

$$\omega(a) \subseteq \omega(a_1) \times \omega(a_2) = \{(0, 0)\} , \tag{5.1.10}$$

However

$$a_1 a_2 + a_2 a_1 = 1 = \begin{pmatrix} 1 & 0 \\ 0 & 1 \end{pmatrix} , \tag{5.1.11}$$

excluding this point. It follows

$$\omega(a) = \emptyset . \tag{5.1.12}$$

With $p = z_1 z_2 + z_2 z_1$ it is clear

$$\emptyset = p\omega(a) \neq \omega p(a) = \{1\} . \tag{5.1.13}$$

There is however equality in (5.1.6) for certain systems p of polynomials and rationals: indeed if

$$q \circ p = z = p \circ q \tag{5.1.14}$$

then (5.1.6) applied to each of p and q guarantees equality in (5.1.6) for the other. For another example suppose that $p \in \text{Poly}_n^m$ with $m = n + k$ takes the form

$$p = (z, q) \tag{5.1.15}$$

with $q \in \text{Poly}_n^k$: note the implication

$$(\lambda, \mu) \in \omega(a, q(a)) \Longrightarrow \mu = q(\lambda) \tag{5.1.16}$$

whenever (5.1.6) is valid.

In the same vein we observe that if $*$ is an involution on A in the sense (1.5.21), satisfying the "hermitian" condition (1.5.28), then

$$(\lambda, \mu) \in \omega(a, a^*) \Longrightarrow \mu = \bar{\lambda} . \tag{5.1.17}$$

There are also *left and right point*, and *approximate point* spectrum for n-tuples of Banach algebra elements: we define, for a ring A,

$$A_{left}^{on} = \{a \in A^n : \bigcap_{j=1}^n L_{a_j}^{-1}(0) = \{0\}\} , \quad A_{right}^{on} = \{a \in A^n : \bigcap_{j=1}^n R_{a_j}^{-1}(0) = \{0\}\} , \tag{5.1.18}$$

and, for a Banach algebra A,

$$\begin{aligned} A_{left}^{\bullet n} &= \{a \in A^n : \inf\{\sum_{j=1}^n \|a_j x\| : \|x\| \geq 1\} > 0\} , \\ A_{right}^{\bullet n} &= \{a \in A^n : \inf\{\sum_{j=1}^n \|x a_j\| : \|x\| \geq 1\} > 0\} \end{aligned} \tag{5.1.19}$$

Evidently

$$A_{left}^{-1} \subseteq A_{left}^{\bullet n} \subseteq A_{left}^{on} , \ A_{right}^{-1} \subseteq A_{right}^{\bullet n} \subseteq A_{right}^{on} . \tag{5.1.20}$$

Now we define, for $a \in A^n$

$$\pi^{left}(a) = \{\lambda \in \mathbb{C}^n : a - \lambda \notin A_{left}^{on}\} , \ \pi^{right}(a) = \{\lambda \in \mathbb{C}^n : a - \lambda \notin A_{right}^{on}\} . \tag{5.1.21}$$

Similarly, in a Banach algebra, we write

$$\tau^{left}(a) = \{\lambda \in \mathbb{C}^n : a - \lambda \notin A_{left}^{\bullet n}\} , \ \tau^{right}(a) = \{\lambda \in \mathbb{C}^n : a - \lambda \notin A_{right}^{\bullet n}\} . \tag{5.1.22}$$

Evidently

$$\pi^{left}(a) \subseteq \tau^{left}(a) \subseteq \sigma^{left}(a) , \pi^{right}(a) \subseteq \tau^{right}(a) \subseteq \sigma^{right}(a) . \tag{5.1.23}$$

The remainder theorem for non commutative polynomials ensures inclusion (5.1.6) whenever, extending (5.1.7),

$$\omega \in \{\pi^{left}, \pi^{right}, \tau^{left}, \tau^{right}\} . \tag{5.1.24}$$

For example, if $A = B(X)$ for a Hilbert space X, then a normal eigenvalue for $a \in A$ is $\lambda \in \mathbb{C}$ for which

$$(\lambda, \bar{\lambda}) \in \pi^{left}(a, a^*) ; \tag{5.1.25}$$

we leave it to the reader to decide what is a "normal approximate eigenvalue".

When A is a C* algebra and $\{a, b\} \subseteq A$ then "topological norm additivity" (4.3.30) gives inequality

$$\|a^*a\| \leq \|a^*a + b^*b\| . \tag{5.1.26}$$

Equality in (5.1.6) in general fails for point and approximate point spectrum as well as for spectrum: however if A is a complex Banach algebra then there is indeed, as we are about to demonstrate, equality in (5.1.6), for both the left and the right spectrum, and indeed also the left and the right approximate point spectrum, for arbitrary systems of polynomials, for *commuting* systems $a = (a_1, a_2, \ldots, a_n) \in A^n$.

5.2   Relative and Restricted Spectrum

As a framework for the argument for the spectral mapping theorem we introduce, with ω satisfying (5.1.6), the *relative and restricted spectrum* of a system $a \in A^n$ with respect to a system $b \in A^m$ by setting, for $\mu \in \mathbb{C}^m$,

$$\omega_{b=\mu}(a) = \{\lambda \in \mathbb{C}^n : (\lambda, \mu) \in \omega(a, b)\} . \tag{5.2.1}$$

Intuitively if $A = C(\Omega)$ is the algebra of continuous functions on a compact Hausdorff space Ω then the spectrum of $\sigma(a)$ of $a \in A$ is the range $a(\Omega) \subseteq \mathbb{C}$ of the function $a \in A$, while an m-tuple $b \in A^m$ can be interpreted as a continuous mapping from Ω to $\mathbb{C}^m$ and again the spectrum $\sigma(b) \subseteq \mathbb{C}^m$ is the range of the mapping. Then the "*restricted spectrum*"

$$\sigma_{a=\lambda}(b) = \{b(t) : a(t) = \lambda\} \subseteq \{b(t) : t \in \Omega\} , \tag{5.2.2}$$

is just the range of the restriction of the mapping $b : \Omega \to \mathbb{C}^m$ to the "level surface" $a^{-1}(\lambda) \subseteq \Omega$. Now the *relative spectrum*

$$\omega_{a=a}(b) = \bigcup_{\lambda \in \mathbb{C}^n} \omega_{a=\lambda}(b) . \tag{5.2.3}$$

is just the union of all the restricted spectra. More prosaically $\omega_{a=a}(b) \subseteq \mathbb{C}^m$ is the image of the joint spectrum $\omega(a, b) \subseteq \mathbb{C}^{m+n}$ under the natural projection $(\lambda, \mu) \mapsto \mu$. As an example if the idempotent $p = p^2 \in A$ commutes with $a \in A$ then, with $B = pAp$, there is equality

$$\sigma_{p=1}^{left}(a) = \sigma_B^{left}(ap) , \tag{5.2.4}$$

$$\sigma_{p=1}^{right}(a) = \sigma_B^{right}(ap) \tag{5.2.5}$$

and hence also, with $\sigma = \sigma^{left} \cup \sigma^{right}$,

$$\sigma_{p=1}(a) = \sigma_B(ap) . \tag{5.2.6}$$

Equality (5.2.4) continues to hold if more generally

$$ap = pap . \tag{5.2.7}$$

When $A = B(X)$ then (5.2.7) means that the range $Y = p(X)$ is an invariant subspace for $a \in A$, and pAp is the algebra of restrictions:

$$Y = p(X) \implies pAp \cong B(Y) . \tag{5.2.8}$$

5.3   The Spectral Mapping Theorem

By (5.1.6) there is inclusion, with $\omega \in \{\sigma^{left}, \sigma^{right}\}$ and more generally, and a $\in$
A^n and $b \in A^m$,

$$\omega(a, b) \subseteq \omega(a) \times \omega(b) \subseteq \mathbb{C}^{m+n} . \tag{5.3.1}$$

Now we claim that if $p \in \text{Poly}_n^m$ then, again by (5.1.6),

$$b = p(a) \Longrightarrow \omega(a) = \omega_{b=b}(a) , \tag{5.3.2}$$

and

$$b = p(a) \Longrightarrow \omega_{a=a}(b) = p\omega(a) . \tag{5.3.3}$$

Now if $c \in A$ commutes with $b = (b_1, b_2, \ldots, b_m) \in A^m$, in the sense that

$$cb_j = b_j c \text{ for each } j \in \{1, 2, \ldots, m\} , \tag{5.3.4}$$

then we claim that, with $\omega = \sigma^{left}$ and then similarly with $\omega = \sigma^{right}$, there is
equality $\omega(b) = \omega_{a=a}(b)$. For if $\mu \in \omega(b) \subseteq \mathbb{C}^m$ then $b - \mu$ generates a proper
left ideal in A, whose closure $N = \text{cl} \sum_{j=1}^m A(b_j - \mu_j)$ also excludes the identity
$1 \in A$. Thus N is a proper closed left ideal of A. By (5.3.4)

$$c \in M = N^{-1}N = \{d \in A : Nd \subseteq N\} , \tag{5.3.5}$$

where $N^{-1}N$ is the *residual quotient* of (1.1.8). Now $1 \in M \subseteq A$ is by (1.1.13) a
closed subalgebra of A and $1 \notin N$ is a proper closed two-sided ideal of M. Thus the
quotient $B = M/N$ is a non trivial Banach algebra. Next, if

$$\lambda \in \partial \sigma_B(a + N) \subseteq \tau_B^{left}(a + N) , \tag{5.3.6}$$

then we claim that

$$(\lambda, \mu) \in \sigma^{left}(a, b) : \tag{5.3.7}$$

for if to the contrary

$$a'(a - \lambda) + b' \cdot (b - \mu) = 1$$

then for arbitrary $x \in M$ we have

$$x = a'(a - \lambda)x + b' \cdot (b - \mu)x \in a'(a - \lambda)x + N ,$$

giving, in violation of (5.3.6),

$$\|x + N\| \le \|a'\| \, \|(a - \lambda)x + N\| \, .$$

It now follows, by induction on $n \in \mathbb{N}$, that if $a = (a_1, a_2, \ldots, a_n) \in A^n$ is commutative and commutes with $b \in A^m$, in the sense that $c = a_i$ satisfies (5.3.4) for each $i = 1, 2, \ldots, n$, then

$$\omega(b) = \omega_{a=a}(b) \, . \tag{5.3.8}$$

In particular (5.3.2) holds with $b = p(a)$.

Generally for arbitrary $a \in A^n$ and $b \in A^m$ and $q \in \mathrm{Poly}_{n+m}^k$ there is by (5.1.6) inclusion

$$q\omega(a, b) \subseteq \omega q(a, b) \, , \tag{5.3.9}$$

while if $a \in A^n$ is commutative and commutes with $b \in A^m$ then there is (remainder theorem again) equality, for arbitrary $\lambda \in \mathbb{C}^n$,

$$\omega_{a=\lambda} g(a, b) = \omega_{a=\lambda} g(\lambda, b) \, . \tag{5.3.10}$$

If instead $b \in A^m$ is commutative and commutes with $a \in A^n$ then

$$\omega_{b=\mu} g(a.b) = \omega_{b=\mu} g(a, \mu) \, . \tag{5.3.11}$$

If the whole system $(a, b) \in A^{m+n}$ is commutative then by what we have just proved, with now $(a, b) \in A^{m+n}$ in place of $a \in A^n$, there is equality

$$\omega q(a, b) = q\omega(a, b) \, . \tag{5.3.12}$$

The intermediate equalities (5.3.10) and (5.3.11) will suggest a certain kind of "vector-valued" spectrum for vector-valued functions and tensor product elements.

Equality (5.3.12), which we have just proved, for the left spectrum $\omega = \sigma^{left}$ is what we mean by "the spectral mapping theorem"; when $b \in A^m$ with $m = 0$ is vacuous, it reduces to equality in (5.1.6) for commuting $a \in A^n$. At the same time we have of course, by "reversal of products", also proved it for $\omega = \sigma^{right}$, and therefore also for the "so-called Harte spectrum" $\sigma^{left} \cup \sigma^{right}$.

A several variable extension of (4.5.5) seems unlikely: if for example $c \in A$ and $(a, b) = (p(c), q(c)) \in A^2$ then

$$\omega(a, b) = \langle p, q \rangle \omega(c) \subseteq \partial\omega(a, b) \, , \tag{5.3.13}$$

being in a sense one dimensional in a two dimensional environment.

5.4   Gelfand's Theorem

It is instructive to compare the argument for (5.3.8) with the proof of *Gelfand's theorem*, which says that if A is a commutative Banach algebra then the *Gelfand mapping*

$$T = \Gamma : A \to B = C(\sigma(A)), \tag{5.4.1}$$

sending

$$a \mapsto a^\wedge : \varphi \mapsto \varphi(a) \tag{5.4.2}$$

from A to the continuous functions on the spectrum

$$\sigma(A) = \{\varphi \in A^* : \varphi(a'a) - \varphi(a')\varphi(a) = 0 = 1 - \varphi(1)\} \tag{5.4.3}$$

has the "Gelfand property" (1.5.6).

Now if $\mu \in \omega(b) = \sigma^{left}(b) = \sigma^{right}(b)$ then $b - \mu$ generates a maximal proper ideal $N \in ML(A) = MR(A)$ of A, and of course $M = A$ and the quotient $B = A/N$ is one dimensional, and hence the spectrum $\sigma_B(a + N) = \{\lambda\}$ is a single point, and the mapping $a \mapsto \lambda$ has the status of a (bounded) *multiplicative linear functional* $\varphi \in \sigma(A)$.

Vladimir Müller has noticed how the spectral mapping theorem in several variables impacts on the spectrum of single elements: if

$$a \mapsto \omega(a)$$

given by a Müller regularity H_ω is the restriction to $n = 1$ of a mapping ω on A^n for which the spectral mapping theorem for polynomials holds then whenever $\{a, b\} \subseteq A$ is commutative there is equivalence

$$ab \in H_\omega \iff \{a, b\} \subseteq H_\omega. \tag{5.4.4}$$

Indeed if $\{a, b\} \subseteq H_\omega \subseteq A$ then

$$\omega(ab) = \{\lambda\mu : (\lambda, \mu) \in \omega(a, b)\} \subseteq \{\lambda\mu : (\lambda, \mu) \in \omega(a) \times \omega(b)\},$$

which does not contain $0 \in \mathbb{C}$. Conversely if for example $a \notin H_\omega$ so that $0 \in \omega(a)$ then there must be $\mu \in \omega(b)$ for which $(0, \mu) \in \omega(a, b)$ and hence

$$0 = 0\mu \in \omega(ab).$$

The condition (5.4.4) is also sufficient for ω to be the tip of a multidimensional iceberg: in the presence of (5.4.4) we may take, for commutative $a \in A^n$,

$$\omega_A(a) = \sigma_D(a) \text{ with } D = \text{comm}^2(a) . \tag{5.4.5}$$

Antoni Wawrzynczyk has, with the aid of Schur's Lemma (1.13.20), obtained an extension of the spectral mapping theorem (5.3.12) to locally convex "Waelbroeck algebras", for which the invertible group is open and the inversion map continuous.

5.5 Joint Approximate Spectrum

To establish the spectral mapping theorem for the approximate point spectrum we have to argue that if $a \in A$ commutes with $b \in A^m$ then there is equality

$$\tau^{left}(b) = \tau^{left}_{a=a}(b) . \tag{5.5.1}$$

As a first step in this direction we observe

$$\pi^{left}(b) \subseteq \tau^{left}_{a=a}(b) . \tag{5.5.2}$$

Indeed if $\mu \in \pi^{left}(b)$ then the subspace

$$N = L^{-1}_{b-\mu}(0) \subseteq A . \tag{5.5.3}$$

is a non zero closed subspace of A with the property

$$ab = ba \implies aN \subseteq N . \tag{5.5.4}$$

Now if we take $\lambda \in \partial\sigma(T)$ where $T : m \mapsto am$ is the restriction of L_a to N then certainly

$$(\lambda, \mu) \in \tau^{left}(a, b) . \tag{5.5.5}$$

To square the circle and convert (5.5.2) to (5.5.1) we will need a piece of machinery: the *enlargement* process of (3.7.4). Indeed if $T = L_b : A \to A^m$ then by (3.7.6)

$$\tau^{left}(b) = \tau^{left}(L_b) = \pi^{left}\mathbf{Q}(L_b) \subseteq \tau^{left}_{a=a}\mathbf{Q}(L_b) = \tau^{left}_{a=a}(b) . \tag{5.5.6}$$

5.6 Tensor Products

If X and Y are linear spaces then the *tensor product*

$$X \otimes Y \equiv X \otimes_{\mathbb{K}} Y \tag{5.6.1}$$

can be interpreted as a certain quotient of the space of all formal linear combinations of the elements of the cartesian product $X \times Y$: specifically by the subspace generated by the set of all such expressions of the following forms:

$$(x_1 + x_2, y) - (x_1, y) - (x_2, y) \; ; \; (x, y_1 + y_2) - (x, y_1) - (x, y_2)$$
$$\lambda(x, y) - (\lambda x, y) \; ; \; \lambda(x, y) = (x, \lambda y) \tag{5.6.2}$$

If we write

$$x \otimes y \tag{5.6.3}$$

for the coset in this quotient which contains the point $(x, y) \in X \times Y$ then the tensor product $X \otimes Y$ is generated by the set of all these *rank one* tensors $x \otimes y$.

For example the tensor product

$$X \otimes_{\mathbb{R}} \mathbb{C}$$

can be thought of as the *complexification* of the real space X.

When X and Y are normed spaces then there are various ways in which the tensor product $X \otimes Y$ can be normed: a *crossnorm* on the tensor product satisfies the condition that, for arbitrary $(x, y) \in X \times Y$,

$$\|x \otimes y\| = \|x\| \, \|y\| \, , \tag{5.6.4}$$

while a *uniform crossnorm* satisfies the additional condition, that

$$\|x \otimes y\| = \sup\{|f(x)| \, |g(y)| \; : \; (f, g) \in X^* \times Y^* \; ; \; \|f\| \, \|g\| \le 1\} . \tag{5.6.5}$$

The fundamental example of a uniform crossnorm is the "projective", or *greatest crossnorm*,

$$\|w\|_1 = \inf\{\sum_{j \in J} \|x_j\| \, \|y_j\| : w = \sum_{j \in J} x_j \otimes y_j\} ; \tag{5.6.6}$$

this most truly represents the essence of a "tensor product". Dually the "inductive", or *least uniform crossnorm*, is given by

$$\|\sum_{j \in J} x_j \otimes y_j\|_\infty = \sup\{|\sum_{j \in J} f(x_j)g(y_j)| : \|f\| \, \|g\| \le 1\} : \tag{5.6.7}$$

here the supremum is taken over $f \in X^*$ and $g \in Y^*$. The normed space $X \otimes Y$ can now be completed, in the greatest crossnorm, to give the Banach space $X \otimes_1 Y$, or in the least uniform crossnorm, to give the Banach space $X \otimes_\infty Y$.

If in particular $X = A$ and $Y = B$ are linear algebras, then the tensor product

$$X \otimes Y = A \otimes B$$

will also be a linear algebra, with multiplication generated linearly by setting

$$(a' \otimes b')(a \otimes b) = a'a \otimes b'b . \tag{5.6.8}$$

When in particular A and B are normed algebras it will be appropriate to look for uniform crossnorms which respect this. The reader may like to check that this is the case for both the greatest, and the least uniform, crossnorm.

When $D = A \otimes B$ is the Banach algebra derived from Banach algebras A and B with such a uniform crossnorm then the spectrum of a rank one tensor (5.6.3) is interesting:

$$\sigma_D(a \otimes b) = \sigma_A(a)\sigma_B(b) . \tag{5.6.9}$$

This is the tip of an iceberg: if $a \in A^n$ and $b \in B^m$ then

$$\sigma_D^{left}(a \otimes 1, 1 \otimes b) = \sigma_A^{left}(a) \times \sigma_B^{left}(b) \subseteq \mathbb{C}^n \times \mathbb{C}^m \tag{5.6.10}$$

and

$$\sigma_D^{right}(a \otimes 1, 1 \otimes b) = \sigma_A^{right}(a) \times \sigma_B^{right}(b) \subseteq \mathbb{C}^n \times \mathbb{C}^m . \tag{5.6.11}$$

For single elements, $n = m = 1$,

$$\partial(\sigma_A(a) \times \sigma_B(b)) \subseteq \sigma_D(a \otimes 1, 1 \otimes b) \subseteq \sigma_A(a) \times \sigma_B(b) \subseteq \mathbb{C}^2 . \tag{5.6.12}$$

To prove this note that it is clear, in each of (5.6.10) and (5.6.11), that the left hand side is included in the right. Conversely, for (5.6.10), suppose that $\lambda \in \sigma_A^{left}(a)$ and $\mu \in \sigma_B^{left}(b)$: then the identity $1 \in A$ is excluded from the left ideal M generated by $a - \lambda$ in A, and the identity $1 \in B$ from the left ideal N generated by $b - \mu$. By the Hahn-Banach theorem there are therefore functionals $f \in A^*$ and $g \in B^*$ for which

$$f(1) = 1 = g(1) ; \quad f(M) = \{0\} = g(N) . \tag{5.6.13}$$

Now

$$M \otimes B + A \otimes N \subseteq D = A \otimes B \tag{5.6.14}$$

is a left ideal which contains each element $(a_j - \lambda_j) \otimes 1$ and each element $1 \otimes (b_k - \mu_k)$: but now

$$(f \otimes g)(1 \otimes 1) = 1 \notin \{0\} = (f \otimes g)(M \otimes B + A \otimes N) . \qquad (5.6.15)$$

For $a \in A$ and $b \in B$ it now follows that for systems of polynomials, for each $\omega \in \{\sigma^{left}, \sigma^{right}\}$,

$$\omega_D g(a \otimes 1, 1 \otimes b) = g(\omega_A(a) \times \omega_B(b)) . \qquad (5.6.16)$$

More generally, if $a \in A^n$ is commutative then, by (5.3.10),

$$\omega_D(g(a \otimes 1, 1 \otimes b) = \omega_B g(\omega_A(a), b) \equiv \bigcap \{\omega_B(\lambda, b) : \lambda \in \omega_A(a)\} ; \qquad (5.6.17)$$

if instead $b \in B^m$ is commutative then, by (5.3.11),

$$\omega_D g(a \otimes 1, 1 \otimes b) = \omega_A g(a, \omega_B(b)) \equiv \bigcap \{\omega_A g(a, \mu) : \mu \in \omega_B(b)\} . \qquad (5.6.18)$$

If A is a ring and M a left and N a right A module then the tensor product

$$N \otimes_A M$$

is a well-defined abelian group: in addition to the material in (5.6.2) include also

$$(na, m) - (n, am) .$$

Now the *mixed identity* tells us what are the abelian group homomorphisms from the tensor product to an arbitrary abelian group P:

$$L(N \otimes_A M, P) \cong L_A(M, L(N, P)) . \qquad (5.6.19)$$

Here of course

$$L_A(M, M') = \{T \in L(M, M') : T(am) = aT(m)\}$$

denotes the "left A linear" mappings from M to M'. For linear spaces which are modules over an algebra, in particular for "normed modules" over a normed algebra, then the normed analogue of the mixed identity is valid provided the tensor product is given the analogue of the "greatest crossnorm":

$$BL(N \otimes_A M, P) \cong BL_A(M, BL(N, P)) . \qquad (5.6.20)$$

Specialising to the case in which P is the scalar field gives an expression for the dual space of a tensor product:

$$(N \otimes_A M)^* \cong BL_A(M, N^*) . \qquad (5.6.21)$$

The mixed identity therefore explains why the real and the complex duals of complex space are (3.6.6) the same, leading to the extension of the Hahn-Banach theorem from real spaces to complex. In the same spirit the true "concrete" interpretation of the rank one tensor $g \otimes x$ is not the operator

$$g \odot x : Y \to X$$

of (1.13.11), but rather the linear functional

$$T \mapsto g(Tx) : L(X, Y) \to \mathbb{K} . \qquad (5.6.22)$$

The same mixed identity also extends from Hilbert to reflexive Banach spaces X that part of the "*von Neumann double commutant theorem*" which says that commutants in $B(X)$ are always dual spaces.

5.7 Elementary Operators

If M is a (left A, right B) bimodule, which means that tuples $a \in A^n$ and $b \in B^n$ combine to give *elementary operators*

$$L_a \circ R_b : x \mapsto \sum_{j=1}^{n} a_j x b_j . \qquad (5.7.1)$$

and if the module M is *prime* , in the sense that there is implication, for arbitrary $(a, b) \in A \times B$,

$$L_a R_b = 0 \in B(M) \Longrightarrow 0 \in \{a, b\} \qquad (5.7.2)$$

then the space

$$L_A \circ R_B = \{L_a \circ R_b : n \in \mathbb{N}, a \in A^n, b \in B^n\} \qquad (5.7.3)$$

closely resembles the tensor product $A \otimes B^{op}$, where B^{op} is derived from B by reversal of product. If A and B are normed algebras and the normed module M is *ultraprime*, in the sense that there is equality

$$\|L_a R_b\| = \|a\| \, \|b\| \qquad (5.7.4)$$

then the operator norm on $L_A \circ R_B$ will also be a uniform crossnorm. Now with

$$D = B(M)$$

we claim that, with $a \in A^n$ and $b \in B^m$,

$$\tau_A^{left}(a) \times \tau_B^{right}(b) \subseteq \sigma_D^{left}(L_a, R_b) \subseteq \sigma_A^{left}(a) \times \sigma_B^{right}(b) , \qquad (5.7.5)$$

and

$$\tau_A^{right}(a) \times \tau_B^{left}(b) \subseteq \sigma_D^{right}(L_a, R_b) \subseteq \sigma_A^{right}(a) \times \sigma_B^{left}(b) . \qquad (5.7.6)$$

For single elements, $n = m = 1$,

$$\partial(\sigma_A(a) \times \sigma_B(b)) \subseteq \sigma_D(L_a, R_b) \subseteq \sigma_A(a) \times \sigma_B(b) . \qquad (5.7.7)$$

The most fundamental of the elementary operators are the *generalized inner derivations* $L_a - R_b$ and the products $L_a R_b$. If $a \in A$ then

$$D = L_a - R_a \Longrightarrow D(xy) = x Dy + (Dx)y . \qquad (5.7.8)$$

More generally $L_a - R_b \in GD(A)$ is a "generalized derivation":

$$D = L_a - R_b \Longrightarrow D(xwy) = D(xw)y - x(Dw)y + xD(wy) . \qquad (5.7.9)$$

Generalized derivations coincide with sums of derivations and multiplications:

$$D(A) + L(A) = GD(A) = D(A) + R(A) , \qquad (5.7.10)$$

where

$$T \in L(A) \Longleftrightarrow T(xy) = (Tx)y ; \ T \in R(A) \Longleftrightarrow T(xy) = x(Ty) .$$

5.8   Quasicommuting Systems

There is an extension of the spectral mapping theorem to n tuples $a \in A^n$ which *quasi commute* in a curious sense. We begin by taking a second look at the derivations of (5.7.8). The *Leibnitz rule* applies to derivations $D \in D(A)$:

$$D^n(xy) = \sum_{r=0}^{n} \binom{n}{r} D^{n-r}(x)D^r(y) . \qquad (5.8.1)$$

The *Kleinecke-Sirokov* theorem and the *Singer-Wermer theorem* together say that bounded derivations breed quasinilpotents: if $D \in D(A) \cap B(A)$ and $a \in A$ satisfy either

$$D^2 a = 0 \qquad (5.8.2)$$

or

$$a(Da) = (Da)a , \qquad (5.8.3)$$

then

$$Da \in QN(A) \qquad (5.8.4)$$

is quasinilpotent. Indeed if $D \in D(A) \cap B(A)$ and $x \in A^n$ satisfy

$$D^2(x_j) = 0 \ (j = 1, 2, \ldots, n)$$

then

$$D^n(x_1 x_2 \ldots x_n) = (n!)(Dx_1)(Dx_2) \ldots (Dx_n) \ and \ D^{n+1}(x_1 x_2 \ldots x_n) = 0 .$$

It follows

$$(n!)\|(Da)^n\| \leq \|D\|^n \|a\|^n . \qquad (5.8.5)$$

The quasinilpotence follows from the observation that

$$(n!)^{-1/n} \to 0 \ (n \to \infty) .$$

We have just established the Kleinecke-Sirokov theorem, which says that (5.8.2) implies (5.8.4). The Singer-Wermer theorem says that also (5.8.3) implies (5.8.4). It was Gerard Murphy who first noticed that this is actually a simple consequence of Kleinecke-Sirokov: just apply it to

$$\Delta_a = L_{Da} - R_{Da} : B(A) \to B(A) , \qquad (5.8.6)$$

and observe that

$$\Delta_a(D) = -L_{Da} ; \implies \Delta_a^2(D) = 0 . \qquad (5.8.7)$$

It would be nice to find a "spectral" proof of Singer-Wermer; in the presence of (5.8.3)

$$\sigma(a) = \sigma_{Da=Da}(a) , \qquad (5.8.8)$$

which we would like to sharpen to

$$\sigma(a) = \sigma_{Da=0}(a) . \qquad (5.8.9)$$

Part of the fallout is the implication

$$Da = 1 \Longrightarrow 1 \in \mathrm{QN}(A) \Longrightarrow A = \{0\} . \qquad (5.8.10)$$

It follows that if, with $(a, b) \in A^2$, $ba - ab$ commutes with either a or b, then $ba - ab$ is quasinilpotent. If in particular $ba - ab$ commutes with both a and b then we shall describe the pair (a, b) as "quasicommutative". More generally we shall say that $a \in A^n$ quasicommutes with $b \in A^m$ provided

$$ba - ab \in A^{mn} \text{ commutes with } (a, b) \in A^{m+n} , \qquad (5.8.11)$$

in the sense that each $b_k a_j - a_j b_k$ commutes with each a_i and with each b_i. Finally, we shall say that $a \in A^n$ is quasicommutative if $aa - aa \in A^{n^2}$ commutes with $a \in A^n$.

To see why the spectral mapping theorem extends to quasicommutative systems we need another definition. Going back to the proof for commuting systems, recall the condition (5.3.5) that

$$N = \sum_{j=1}^m A(b_j - \mu_j) \Longrightarrow Nc \subseteq N :$$

if this holds for arbitrary $\mu \in \mathbb{C}^m$ we shall say that $b \in A^m$ is *completely left invariant* under $c \in A$. If this holds for $c = a_j$ for each j then we shall say that b is completely invariant under $a \in A^n$. Now the proof of the left spectral mapping theorem falls into two parts: we claim that if $ba - ab \in A^{mn}$ commutes with $a \in A^n$ then

$$(b, ba - ab) \in A^{m+mn} \text{ is completely left invariant under } a \in A^n ;$$

if in particular $ba - ab$ commutes with (a, b) then for arbitrary systems $h \in \mathrm{Poly}_m^p$ of polynomials

$$(h(b), ba - ab) \text{ is completely left invariant under } a .$$

If even more particularly $ba - ab$ commutes with (a, b) and a is commutative then

$$(g(a, b), ba - ab) \text{ is completely left invariant under } a .$$

Now, with $\omega = \sigma^{left}$ we find that whenever $ba - ab$ is commutative and commutes with b there is equality

$$\omega(b) = \omega_{ba-ab=0}(b) .$$

Also, whenever a is quasicommutative and commutes with $ba - ab$, we find that

$$\omega_{ba-ab=0}(b) = \omega_{a=a}(b) .$$

Combining these two, it follows that if $a \in A^n$ is quasicommutative and quasicommutes with $b \in A^m$, then

$$\omega(b) = \omega_{a=a}(b) .$$

This of course is now very nearly the spectral mapping theorem: all we have left is to "substitute $b = f(a)$". Not quite as simple as it seems: we first need to prove that whenever $a \in A^n$ is quasicommutative, then it also quasicommutes with $f(a) \in A^m$ for polynomial systems $f \in \text{Poly}_n^m$.

Notice that (5.8.11), for single elements $a, b \in A$ gives, for the inner derivation $D = L_a - R_a$, implication

$$a(Db) = (Db)a \implies Db \notin A^{-1} , \tag{5.8.12}$$

and hence the implication (5.8.10).

5.9 Holomorphic Left Inverses

If Ω is a compact Hausdorff space and $f \in A = C(\Omega, B)$ is a vector-valued continuous function then the spectrum of f is what it ought to be:

$$\sigma_A(f) = \text{cl} \bigcup \{\sigma_B f(t) : t \in \Omega\} . \tag{5.9.1}$$

Allen's theorem says, for holomorphic functions $f \in \text{Holo}(\Omega, A)$, that if $K \subseteq \mathbb{C}$ is compact and $K \subseteq \text{int } \Omega$ then there is implication

$$\{f(t) : t \in K\} \subseteq A_{left}^{-1} \implies f \in \text{Holo}(K, A)_{left}^{-1} \tag{5.9.2}$$

and

$$\{f(t) : t \in K\} \subseteq A_{right}^{-1} \implies f \in \text{Holo}(K, A)_{right}^{-1} . \tag{5.9.3}$$

To see this observe

$$D = \mathrm{cl}\,\mathrm{Holo}(K, A) = A \otimes_\infty \mathrm{Holo}(K) \subseteq C(K, A) : \tag{5.9.4}$$

for if $t \in K$ then

$$f(t) = \frac{1}{2\pi i} \oint_K f(z)(z - t)^{-1} dz = \lim \frac{1}{2\pi i} \sum_{j=1}^{n} f(s_j)(s_j - s_{j-1})(s_j - t)^{-1} .$$

giving

$$f = \lim \sum_{j=1}^{n} b_j \odot a_j \tag{5.9.5}$$

with

$$b_j = (s_j - z)^{-1}, \quad a_j = \frac{1}{2\pi i} f(s_j)(s_j - s_{j-1}) .$$

But now, by (5.6.18), with $B = \mathrm{Holo}(K)$ and $\omega = \sigma^{left}$,

$$\sigma_D^{left}(f) = \bigcap \{\sigma_A^{left} f(t) : t \in K\} , \tag{5.9.6}$$

and dually

$$\{f(t) : t \in K\} \subseteq A_{left}^{-1} \Longrightarrow f \in D_{left}^{-1} \tag{5.9.7}$$

and

$$\{f(t) : t \in K\} \subseteq A_{right}^{-1} \Longrightarrow f \in D_{right}^{-1} . \tag{5.9.8}$$

It is tempting to enquire whether if

$$1 \in \bigcap_{t \in K} Af(t) + f(t)A \subseteq A , \tag{5.9.9}$$

there are also $g \in D$ and $h \in D$ for which

$$1 = gf + fh \in D . \tag{5.9.10}$$

If we were to try and adapt the argument for (5.9.6), we would expect to extend (5.6.18) and (5.6.19) to another "spectrum" ω, where

$$\omega_A(a) = \{\lambda \in \mathbb{C} : 1 \notin A(a - \lambda) + (a - \lambda)A\} . \tag{5.9.11}$$

Unfortunately however the spectral mapping theorem (5.3.12) is unlikely to hold for ω: with $(e, f) \in A^2$ as in (5.1.9),

$$\omega_A(e) = \omega_A(f) = \emptyset . \tag{5.9.12}$$

It is equally unlikely that we can show that implication

$$t \in K \implies f(t) = f(t)g(t)f(t) \in A^\cap , \tag{5.9.13}$$

will be sufficient for there to be $g \in D$ for which

$$f = fgf \in D^\cap. \tag{5.9.14}$$

Indeed if (5.9.14) holds, the implication (4.8.13) for Kato invertibility suggests that f would then be "hyperexact" on K.

Recall however an observation of Hochwald and Morell, which says that when a ring A is "centre-commutative", in the sense that

$$A = \operatorname{comm}(A) + A^{-1} , \tag{5.9.15}$$

then if, on a semigroup K for which $A^{-1} \subseteq K \subseteq A^\cap$, the mapping $g : K \to A$ satisfies, for every $a \in K$,

$$a = ag(a)a \in A^\cap$$

and also

$$g(\operatorname{comm}^{-1}(a)) \subseteq \operatorname{comm} g(a)$$

then it follows

$$g(K) \subseteq A^\cup \tag{5.9.16}$$

decomposably regular.

5.10 Operator Matrices

If A is a complex linear algebra and $q \in \mathbb{N}$ then "$q \times q$ matrices with entries in A" can be recognised as a tensor product:

$$A^{q \times q} \cong A \otimes \mathbb{C}^{q \times q} : \tag{5.10.1}$$

At the same time, provided the entries (a_{ij}) of $a \in A^{q \times q}$ mutually commute, it is possible to repeat almost verbatim the development of a *determinant*

$$\det(a) = \sum \{\operatorname{sgn}(\pi) a_{1\pi(1)} a_{2\pi(2)} \ldots a_{q\pi(q)} : \pi \in \operatorname{Perm}(1, 2, \ldots, q)\} \in A \,.$$
(5.10.2)

Evidently we can interpret

$$\det(a) = p(a^{\wedge})$$
(5.10.3)

as a polynomial in the tuple

$$a^{\wedge} = (a_{11}, a_{12}, \ldots, a_{1q}, \ldots, a_{qq}) \in A^{q^2} \,.$$
(5.10.4)

Now, taking the obvious basis

$$e = (e_{11}, e_{12}, \ldots, e_{1q}, \ldots, e_{qq}) \in B^{q^2} \,,$$
(5.10.5)

for the numerical matrices $B = \mathbb{C}^{q \times q}$, we are able to write

$$a = g(a^{\wedge}, e) = \sum_{i=1}^{q} \sum_{j=1}^{q} a_{ij} \otimes e_{ij} \in A \otimes B = D :$$
(5.10.6)

now

$$\sigma_D^{left}(a) = g(\sigma_A^{left}(a), e) = \bigcup \{ \sum_{i=1}^{q} \sum_{j=1}^{q} \lambda_{ij} e_{ij} : \lambda \in \sigma_A^{left}(a^{\wedge})\}$$
(5.10.7)

and

$$\sigma_D^{right}(a) = g(\sigma_A^{right}(a), e) = \bigcup \{ \sum_{i=1}^{q} \sum_{j=1}^{q} \lambda_{ij} e_{ij} : \lambda \in \sigma_A^{right}(a^{\wedge})\} \,,$$
(5.10.8)

and hence, for each $\omega \in \{\sigma^{left}, \sigma^{right}, \sigma\}$,

$$\omega_D(a) = \{\lambda \in \mathbb{C} : 0 \in \omega_A \det(a - \lambda)\} \,.$$
(5.10.9)

In a sense—taking our cue from the idea of "partitioned matrices"—the case $q = 2$ is fundamental. If A and B are rings and if M and N are "(A, B) bimodules" then, as in (1.11.2),

$$G = \begin{pmatrix} A & M \\ N & B \end{pmatrix} = \{ \begin{pmatrix} a & m \\ n & b \end{pmatrix} : (a, m, n, b) \in A \times M \times N \times B\}$$

is again a ring; indeed what we mean by "(A, B) bimodules" is visible when we multiply these objects by the usual rules for 2×2 matrices. For example M satisfies the condition (5.7.1). Two particular cases are of interest: the "commutative" case, in which

$$A = M = N = B \text{ and } \{a, m, n, b\} \text{ is commutative} , \tag{5.10.10}$$

and the "triangular" case, in which

$$\text{either } N = \{0\} \text{ or } M = \{0\} : \tag{5.10.11}$$

more generally just assume either $n = 0$ or $m = 0$.
    We remark that if

$$T = \begin{pmatrix} a & m \\ 0 & b \end{pmatrix} \in G = \begin{pmatrix} A & M \\ N & B \end{pmatrix} \tag{5.10.12}$$

and if $a \mapsto a^{\sim}$ and $b \mapsto b^{\sim}$ are adjugate in the sense of (1.11.16) then

$$T \mapsto \text{adj}(T) = T^{\sim} = \begin{pmatrix} |b|a^{\sim} & -a^{\sim}mb^{\sim} \\ 0 & |a|b^{\sim} \end{pmatrix} , \tag{5.10.13}$$

is again a adjugate, with

$$\det(T) = |T| = |a| \, |b| . \tag{5.10.14}$$

Similarly if

$$a = aa'a , \quad b = bb'b , \quad m = (aa')m(b'b)$$

then

$$T = \begin{pmatrix} a & m \\ 0 & b \end{pmatrix} , \quad T' = \begin{pmatrix} a' & -a'mb' \\ 0 & b' \end{pmatrix}$$

gives

$$T = TT'T .$$

Evidently, in the situation of (5.10.12), invertibility $a \in A^{-1}$, $b \in B^{-1}$ and invertibility $T \subset G^{-1}$ conform to the democratic consensus (1.2.17).
    If $(a, b, c, u, v) \in A^5$ satisfy

$$vb = bu \tag{5.10.15}$$

then there is (*Slodkowski's Lemma*) implication

$$(c, b) \; exact \; and \; (b, a) \; exact \Longrightarrow (\begin{pmatrix} c & 0 \\ v & b \end{pmatrix}, \begin{pmatrix} b & 0 \\ -u & a \end{pmatrix}) \; exact. \qquad (5.10.16)$$

Conversely

$$(\begin{pmatrix} c & 0 \\ v & b \end{pmatrix}, \begin{pmatrix} b & 0 \\ -u & a \end{pmatrix}) \; exact \Longrightarrow (\begin{pmatrix} c \\ v \end{pmatrix}, b) \; exact \; and \; (b, (a \; u)) \; exact.$$

$$(5.10.17)$$

For (5.10.16) argue that, with $c \in C^k$, $b \in B^k$ and $a \in A^k$ then, in the notation of (5.1.3), if $c' \cdot c + b \cdot b'' = 1 = b' \cdot b + a \cdot a'$ then

$$\begin{pmatrix} c' & 0 \\ -b'vc' & b' \end{pmatrix} \cdot \begin{pmatrix} c & 0 \\ v & b \end{pmatrix} + \begin{pmatrix} b & 0 \\ -u & a \end{pmatrix} \cdot \begin{pmatrix} b'' & 0 \\ a'ub'' & a' \end{pmatrix} = \begin{pmatrix} 1 & 0 \\ 0 & 1 \end{pmatrix}; \qquad (5.10.18)$$

conversely, for (5.10.17), if

$$\begin{pmatrix} c' & w' \\ v' & b' \end{pmatrix} \cdot \begin{pmatrix} c & 0 \\ v & b \end{pmatrix} + \begin{pmatrix} b & 0 \\ -u & a \end{pmatrix} \cdot \begin{pmatrix} b'' & w'' \\ -u' & a' \end{pmatrix} = \begin{pmatrix} 1 & 0 \\ 0 & 1 \end{pmatrix} \qquad (5.10.19)$$

then

$$(c' \; w') \cdot \begin{pmatrix} c \\ v \end{pmatrix} + b \cdot b'' = 1 = b' \cdot b + (-u \; a) \cdot \begin{pmatrix} w'' \\ a' \end{pmatrix}. \qquad (5.10.20)$$

When in particular $k = 1$ and $A = L(X)$, $B = L(Y)$ and $C = L(Z)$ are linear operators then there are also non-splitting versions of (5.10.16) and (5.10.17):

$$c^{-1}(0) \times b^{-1}(0) \subseteq b(Y) \times a(X) \Longrightarrow \begin{pmatrix} c & 0 \\ v & b \end{pmatrix}^{-1} \begin{pmatrix} 0 \\ 0 \end{pmatrix} \subseteq \begin{pmatrix} b & 0 \\ -u & a \end{pmatrix} \begin{pmatrix} Y \\ X \end{pmatrix}.$$

$$(5.10.21)$$

and conversely

$$\begin{pmatrix} c & 0 \\ v & b \end{pmatrix}^{-1} \begin{pmatrix} 0 \\ 0 \end{pmatrix} \subseteq \begin{pmatrix} b & 0 \\ -u & a \end{pmatrix} \begin{pmatrix} Y \\ X \end{pmatrix} \Longrightarrow \begin{pmatrix} c \\ v \end{pmatrix}^{-1} \begin{pmatrix} 0 \\ 0 \end{pmatrix} \times b^{-1}(0) \subseteq b(Y) \times (-u \; a) \begin{pmatrix} Y \\ X \end{pmatrix}.$$

$$(5.10.22)$$

We may also recall here the "middle non singularity" condition (1.14.13) for a pair (S, T) of linear operators on a space X.

5.11 Spectral Disjointness

If

$$T = \begin{pmatrix} a & m \\ n & b \end{pmatrix} \in \begin{pmatrix} A & M \\ N & B \end{pmatrix} = G \qquad (5.11.1)$$

and

$$Q = Q^2 = \begin{pmatrix} 1 & 0 \\ 0 & 0 \end{pmatrix} \qquad (5.11.2)$$

then, recalling (1.11.7), T commutes with Q if and only if it is (block) diagonal:

$$TQ = QT \iff m = 0 = n . \qquad (5.11.3)$$

In this situation

$$\sigma_A(a) \cap \sigma_B(b) = \emptyset \iff Q \in \text{Holo}(T) . \qquad (5.11.4)$$

To see this note that

$$f(T) = Q \implies (f(a) = 1 \in A \ and \ f(b) = 0 \in B) , \qquad (5.11.5)$$

giving

$$\sigma_A(a) \cap \sigma_B(b) \subseteq f^{-1}(1) \cap f^{-1}(0) = \emptyset . \qquad (5.11.6)$$

Conversely if a and b have disjoint spectra then $Q = \delta_K(T)$ where

$$K = \sigma_A(a) \subseteq \sigma_G(T), \qquad (5.11.7)$$

and δ_K is the *characteristic function* (3.8.5) of $K \subseteq \mathbb{C}$, hence

$$\delta_K(T) = \frac{1}{2\pi i} \oint_K (zI - T)^{-1} dz , \qquad (5.11.8)$$

the Cauchy integral formula (4.3.7).

If more generally

$$\sigma_A^{left}(a) \cap \sigma_B^{right}(b) - \sigma_A^{right}(a) \cap \sigma_B^{left}(b) - \emptyset \qquad (5.11.9)$$

then

$$Q \in \text{comm}_G^2(T) . \qquad (5.11.10)$$

Indeed (5.11.9) says, applying the spectral mapping theorem (5.3.11) to each of the commuting pairs $(L_a, R_b) \in B(M)^2$ and $(L_b, R_a) \in B(N)^2$, both $L_a - R_b \in B(M)_{left}^{-1}$ and $L_b - R_a \in B(N)_{left}^{-1}$ are left invertible, hence in particular one-one, giving

$$\text{comm}_G(T) \subseteq \text{comm}_G(Q) , \tag{5.11.11}$$

which is the same as (5.11.10). At the same time (5.11.9) also says that $L_a - R_b$ and $L_b - R_a$ are each right invertible, hence in particular onto, giving an extension of the splitting exactness (1.10.3) to (b, a) and (a, b):

$$M \subseteq aM + Mb , \quad N \subseteq Na + bN . \tag{5.11.12}$$

(5.11.9) therefore says that $L_a - R_b \in B(M)^{-1}$ is invertible, so that the "Sylvester-Rosenblum equation"

$$ax - xb = y$$

always has a unique solution. In another direction, strengthened versions of (5.11.4) contribute to more detailed constructions for $(L_a - R_b)^{-1} \in B(M)$. For example if $b \in B^{-1}$ then

$$|a|_\sigma |b^{-1}|_\sigma < 1 \Longrightarrow (L_a - R_b)^{-1} = \sum_{n=0}^{\infty} L_a^n R_b^{-n-1} .$$

Necessary and sufficient for there to be a polynomial p for which

$$|p(a)|_\sigma |b^{-1}|_\sigma < 1$$

is, strengthening the left hand side of (5.11.4),

$$\eta \sigma_A(a) \cap \sigma_B(b) = \emptyset .$$

Another side effect, in the special case $A = B$, of spectral disjointness (5.11.9) is that, with $D_c = L_c - R_c$, there is equality

$$D_{a+b}^{-1}(0) \cap D_{ab}^{-1}(0) = D_a^{-1}(0) \cap D_b^{-1}(0) ,$$

and dually

$$D_{a+b}(A) + D_{ba}(A) = D_a(A) + D_b(A) .$$

We remark also that if $\sigma_A(a) \cap \sigma_B(b) = \emptyset$ then, by (1.11.10),

$$\sigma_G(T) = \sigma_A(a) \cup \sigma_B(b) . \tag{5.11.13}$$

If $T \in G$ is an upper spectral triangle then, with no assumption of spectral disjointness,

$$\sigma_A^{left}(a) \subseteq \sigma_G^{left}(T) \subseteq \sigma_A^{left}(a) \cup \sigma_B^{left}(b) ; \tag{5.11.14}$$

$$\sigma_B^{right}(b) \subseteq \sigma_G^{right}(T) \subseteq \sigma_A^{right}(a) \cup \sigma_B^{right}(b) ; \tag{5.11.15}$$

$$\sigma_B^{left}(b) \subseteq \sigma_G^{left}(T) \cup \sigma_A^{right}(a) ; \tag{5.11.16}$$

$$\sigma_A^{right}(a) \subseteq \sigma_G^{right}(T) \cup \sigma_B^{left}(B) . \tag{5.11.17}$$

It follows

$$\sigma_A^{left}(a) \cup \sigma_B^{right}(b) \subseteq \sigma_G(T) \subseteq \sigma_A(a) \cup \sigma_B(b) \\ \subseteq \sigma_G(T) \cup (\sigma_A^{right}(a) \cap \sigma_B^{left}(b)) \tag{5.11.18}$$

Further

$$\partial(\sigma_A(a) \cup \sigma_B(b)) \subseteq \sigma_G(T) \subseteq \sigma_A(a) \cup \sigma_B(b) \subseteq \eta\sigma_G(T) , \tag{5.11.19}$$

and also

$$\partial(\sigma_A(a) \cup \sigma_B(b)) \subseteq \partial\sigma_A(a) \cup \sigma_B(b) \subseteq \partial\sigma_G(T) . \tag{5.11.20}$$

With spectral disjointness (5.11.4) we also have by (4.6.19) the analogue of (5.11.13) for the exponential spectrum:

$$\varepsilon_G(T) = \varepsilon_A(a) \cup \varepsilon_B(b) . \tag{5.11.21}$$

5.12 Joint Local Spectrum

If X is a Banach space and $T = (T_1, T_2, \ldots, T_n) \in B(X)^n$ then an analogue of the transfinite range (4.8.5) would be

$$R_\omega(T) = \{x \equiv (T - zI) \cdot f(z) : f \in \text{Holo}(0, X^n)\} \subseteq X , \tag{5.12.1}$$

containing $x \in X$ if and only if there is $f \in \mathrm{Holo}(0, X^n)$ for which, near $z = 0$,

$$\sum_{j=1}^{n} (T_j - z_j I) f_j(z) \equiv x \tag{5.12.2}$$

Equivalently, writing $(j = 1, 2, \ldots, n)$,

$$f_j(z) \equiv \sum_{r=0}^{\infty} \sum_{|k|=r} z^k \xi_k^j , \tag{5.12.3}$$

with $\xi^j \in k_\infty(X)$,

$$\sum_{j=1}^{n} T_j \xi_0^j = x , \tag{5.12.4}$$

and, with $z_j z^{k/j} = z^k$,

$$\sum_{j=1}^{n} \sum_{|k|=1}^{\infty} z^k (T_j \xi_k^j - \xi_{k/j}^j) = 0 . \tag{5.12.5}$$

The analogue of the holomorphic kernel would be given by the intersection

$$N_\omega(T) = R_\omega(T) \cap \bigcap_{j=1}^{n} T_j^{-1}(0) . \tag{5.12.6}$$

For a multidimensional version of "SVEP" and the local eigenvalues of (4.8.12) we have

$$\lambda \in \pi_{loc}(T) \iff N_\omega(T - \lambda I) \neq \{0\} . \tag{5.12.7}$$

There can however be no multidimensional version of Kato invertibility: Vladimir Müller has an example which shows that (5.4.4) fails here.

5.13 Taylor Invertibility

Suppose $z = (z_1, z_2, \ldots, z_n)$ and write

$$\mathbb{C}_\Lambda(dz) = \sum_{k=0}^{n} \sum_{|j|=k} \mathbb{C} d_j z \tag{5.13.1}$$

for the exterior algebra of complex differential forms: here for example we write

$$d_{ij}z = dz_i \wedge dz_j = -dz_j \wedge dz_i \tag{5.13.2}$$

whenever

$$\{i, j\} \subseteq \{1, 2, \ldots, n\} ;$$

it follows that

$$dz_j \wedge dz_j = 0 \ (j = 1, 2, \ldots, n) .$$

If X is a complex vector space then the tensor product

$$X_\Lambda(dz) = X \otimes \mathbb{C}_\Lambda(dz) \tag{5.13.3}$$

consists of "vector-valued", or X-valued forms in z. Now an n tuple $T = (T_1, T_2, \ldots, T_n) \in L(X)^n$ induces a linear operator

$$\mathbf{T} = T_\Lambda(dz) : \sum_{k=0}^{n} \sum_{|j|=k} x_j d_j z \mapsto \sum_{i=1}^{n} \sum_{k=0}^{n} \sum_{|j|=k} T_i x_j \otimes dz_i \wedge d_j z \tag{5.13.4}$$

on the space

$$\mathbf{X} = X_\Lambda(dz) .$$

Notice

$$T \in L(X)^n \ \textit{commutative} \iff \mathbf{T}^2 = \mathbf{O} \in L(\mathbf{X}) . \tag{5.13.5}$$

Now, whether or not T is commutative, we shall declare it to be *Taylor non singular* when the pair $(\mathbf{T}, \mathbf{T})$ is (linearly) exact, in the sense

$$\mathbf{T}^{-1}(0) \subseteq \mathbf{T}(\mathbf{X}) , \tag{5.13.6}$$

and to be *Taylor invertible* when the pair $(\mathbf{T}, \mathbf{T})$ is splitting exact, in the sense that there are $\mathbf{T}'$ and $\mathbf{T}''$ in $B(\mathbf{X})$ for which

$$\mathbf{T}''\mathbf{T} + \mathbf{T}\mathbf{T}' = \mathbf{I} \in B(\mathbf{X}) . \tag{5.13.7}$$

For example (1.14.13) if $n = 2$ then a commuting pair $T = (T_1, T_2) \in B(X)^2$ is Taylor non singular provided

$$T_1^{-1}(0) \cap T_2^{-1}(0) = \{0\} \; ; \tag{5.13.8}$$

$$T_1(X) + T_2(X) = X \; ; \tag{5.13.9}$$

and also there is implication, for arbitrary $(x_1, x_2) \in X^2$,

$$T_1 x_2 = T_2 x_1 \implies \exists x_0 \in X \; : \; (x_1, x_2) = (T_1 x_0, T_2 x_0) \in X^2 \; . \tag{5.13.10}$$

When A is a general ring, and $a = (a_1, a_2, \ldots, a_n) \in A^n$ then the following four conditions are equivalent:

$$L_a \text{ is Taylor non singular} \; ; \tag{5.13.11}$$

$$R_a \text{ is Taylor non singular} \; ; \tag{5.13.12}$$

$$L_a \text{ is Taylor invertible} \; ; \tag{5.13.13}$$

$$R_a \text{ is Taylor invertible} \; . \tag{5.13.14}$$

When one, and hence all, of these four conditions holds we shall say that $a \in A^n$ is *Taylor invertible in* A.

In an alternative inductive formulation, suppose $a \in A$ is a ring element: then necessary and sufficient for $a \in A^{-1}$ to have a two-sided inverse is that

$$\Lambda_a = \begin{pmatrix} 0 & 0 \\ a & 0 \end{pmatrix} \tag{5.13.15}$$

is (splitting) self-exact in the sense of the various kinds of self exact in the sense of the last paragraph of Sect. 1.10. This happens if and only if the same is true of

$$\Lambda^a = \begin{pmatrix} 0 & a \\ 0 & 0 \end{pmatrix} . \tag{5.13.16}$$

More generally, if a, a', a'' are in A^k then, in the notation (5.1.3),

$$a' \cdot a = 1 = a \cdot a'' \iff \begin{pmatrix} 0 & a' \\ 0 & 0 \end{pmatrix} \cdot \begin{pmatrix} 0 & 0 \\ a & 0 \end{pmatrix} + \begin{pmatrix} 0 & 0 \\ a & 0 \end{pmatrix} \cdot \begin{pmatrix} 0 & a'' \\ 0 & 0 \end{pmatrix} = \begin{pmatrix} 1 & 0 \\ 0 & 1 \end{pmatrix} . \tag{5.13.17}$$

Inductively if $b \in A^n$ is an n-tuple and $c \in A$ then for the $(n+1)$-tuple $a = (b, c)$ we shall define

$$\Lambda_a = \begin{pmatrix} \Lambda_b & O \\ \Delta_c & -\Lambda_b \end{pmatrix} \in \Lambda_A^{n+1} = (\Lambda_A^n)^{2\times 2}, \quad \Lambda^a = \begin{pmatrix} \Lambda^b & \Delta_c \\ O & -\Lambda^b \end{pmatrix}, \quad (5.13.18)$$

where Λ_b and Λ^b in Λ_A^n have already been defined, and $\Delta_c \in \Lambda_A^n$ is the block diagonal generated by $c \in A$. If $a = (b, c) \in A^n \times A$ is commutative we have implication

$$U \Lambda_b + \Lambda_b V = I \in \Lambda_A^n \Longrightarrow$$

$$\begin{pmatrix} U & O \\ U\Delta_c U & -U \end{pmatrix} \Lambda_a + \Lambda_a \begin{pmatrix} V & O \\ V\Delta_c V & -V \end{pmatrix} = I \in \Lambda_A^{n+1}. \quad (5.13.19)$$

If $a \in A^n$ and $p \in \text{Poly}_n^m$ and $\lambda \in \mathbb{C}^n$ then by the remainder theorem (5.1.8) there is inclusion

$$A^m \cdot (p(a) - p(\lambda)) + (p(a) - p(\lambda)) \cdot A^m \subseteq A^n \cdot (a - \lambda) + (a - \lambda) \cdot A^n \subseteq A. \quad (5.13.20)$$

Now the "Taylor invertibility" of $a \in A^n$ is to be given inductively, by setting

$$a \text{ Taylor invertible} \Longleftrightarrow \Lambda_a \text{ splitting self-exact}, \quad\quad (5.13.21)$$

and for operators

$$a \in B(X)^n \text{ Taylor non-singular} \Longleftrightarrow \Lambda_a \text{ self exact}. \quad (5.13.22)$$

Thus (5.13.20) gives the forward spectral mapping theorem for the Taylor split spectrum, at least for commuting systems. For the two-way theorem we follow the left spectrum argument of (5.3.5), replacing the residual quotient $N^{-1}N$ of (1.1.8) associated with the left ideal $N \subseteq A$ by the two-sided quotient $N : N$ of (1.7.12) associated with $N = \Lambda_A^n \Lambda_a + \Lambda_a \Lambda_A^n \subseteq \Lambda_A^n$.

We shall write, for linear operator tuples $a \in A^n$, where $A = L(X)$, $\tau^{Taylor}(a)$, for the spectrum derived from the self-exactness of Λ_a, and more generally $\sigma^{Taylor}(a)$, from the splitting self-exactness of Λ_a. In particular, when $n = 2$, we have

$$\sigma^{Taylor}(a) = \sigma^{left}(a) \cup \sigma^{middle}(a) \cup \sigma^{right}(a). \quad (5.13.23)$$

Notice also, in a C* algebra A, the implication, for $\{b, c\} \subseteq A$

$$1 \in bA + Ac \Longrightarrow b^*b + cc^* \in A^{-1}. \quad (5.13.24)$$

The splitting exactness of the pair (c, b) says that there is $k > 0$ for which, for arbitrary $u \in A$,

$$k\|u\|^2 \leq \|u^*\| \, \|cu\| + \|u^*u\| \, \|u\| \leq \sqrt{2}\|u\|(\|cu\|^2 + \|b^*u\|^2)^{1/2} \, ,$$

and hence

$$k^2\|u\|^2 \leq 4\|u\| \, \|(c^*c + bb^*)u\| \, .$$

Curto has observed (cf (5.1.26)), for $\{b, c\} \subseteq A$, that if

$$b^*b + cc^* = h^{-1} \in A^{-1} \, ; \tag{5.13.25}$$

$$bb^* + c^*c = k^{-1} \in A^{-1} \, ; \tag{5.13.26}$$

$$1 \in Ab + Ac \, ; \tag{5.13.27}$$

$$1 \in bA + cA \, ; \tag{5.13.28}$$

then necessary and sufficient for

$$1 \in bA + Ac \tag{5.13.29}$$

is the condition

$$bhc^* = c^*kb \, . \tag{5.13.30}$$

5.14   Fredholm, Weyl and Browder Theory

If $T \in B(X)$ and $S \in B(Y)$ then the spectrum of

$$S \otimes T \in B(Y) \otimes B(X) \subseteq B(Y \otimes X) \tag{5.14.1}$$

is, according to (5.6.9), given by

$$\sigma(S \otimes T) = \sigma(S)\sigma(T) \, . \tag{5.14.2}$$

For the (Fredholm) *essential spectrum* we have inclusion

$$\sigma_{ess}(S \otimes T) \subseteq \sigma_{ess}(S)\sigma(T) \cup \sigma(S)\sigma_{ess}(T) \, . \tag{5.14.3}$$

The *index* of the operator $S \otimes T - \nu I$, for

$$\nu \in \sigma(S)\sigma(T) \setminus \big(\sigma_{ess}(S)\sigma(T) \cup \sigma(S)\sigma_{ess}(T)\big) \, , \tag{5.14.4}$$

is given by the formula

$$\sum_{j \in J} \text{index}(S - \mu_j I) \dim(T - \lambda_j I)^{-\infty}(0)$$
$$- \sum_{j \in J} \dim(S - \mu_j I)^{-\infty}(0)\text{index}(T - \lambda_j I) \tag{5.14.5}$$

where

$$\{(\mu, \lambda) \in \sigma(S) \times \sigma(T) : \mu\lambda = \nu\} = \{(\mu_j, \lambda_j) : j \in J\} . \tag{5.14.6}$$

It now follows that the Weyl spectrum has the same property as the Fredholm spectrum:

$$\omega_{ess}(S \otimes T) \subseteq \omega_{ess}(S)\sigma(T) \cup \sigma(S)\omega_{ess}(T) . \tag{5.14.7}$$

In general we do not get equality in either (5.14.3) or (5.14.7). This however does occur for the Browder essential spectrum: it turns out

$$\beta_{ess}(S \otimes T) = \beta_{ess}(S)\sigma(T) \cup \sigma(S)\beta_{ess}(T) . \tag{5.14.8}$$

Further, if it should happen that

$$\beta_{ess}(S) = \omega_{ess}(S) \text{ and } \beta_{ess}(T) = \omega_{ess}(T) , \tag{5.14.9}$$

then it is necessary and sufficient, for equality in (5.14.7), that

$$\beta_{ess}(S \otimes T) = \omega_{ess}(S \otimes T) . \tag{5.14.10}$$

A tuple of operators $T = (T_1, T_2, \ldots, T_n) \in B(X)^n$ can be described as *left Fredholm*, or *right Fredholm*, if the tuple

$$a = T + B_0(X) \in A^n$$

of cosets is left, or right, invertible in the Calkin algebra $A = B(X)/B_0(X)$: thus for example the spectral mapping theorem for the *left and right essential spectrum* follow from the spectral mapping theorem for Banach algebra elements, for tuples T of operators which commute modulo the compact operators.

There is also a "Taylor Fredholm theory": we can ask, for a commuting system $T \in B(X)^n$, that (5.13.6) be relaxed to

$$\dim \mathbf{T}^{-1}(\mathbf{O})/\mathbf{T}(\mathbf{X}) < \infty , \tag{5.14.11}$$

or assume the analogue of (5.13.7) in the Calkin algebra. The "index" of a Taylor-Fredholm tuple turns out to be given by the *Euler number* of the associated Koszul complex. This however is not visible the formulation (5.13.11): if we instead break the operator $\mathbf{T}=\mathbf{R}+\mathbf{S}$ of (5.13.4) into the sums of "odd" and "even" terms then

$$\mathbf{RS} = \mathbf{O} = \mathbf{SR} \,, \tag{5.14.12}$$

with

$$\max\left(\dim\ \mathbf{S}^{-1}(\mathbf{O})/\mathbf{R}(\mathbf{X}),\, \dim\ \mathbf{R}^{-1}(\mathbf{O})/\mathbf{S}(\mathbf{X})\right) < \infty \,, \tag{5.14.13}$$

and now

$$\mathrm{Index}(T) = \dim\ \mathbf{S}^{-1}(\mathbf{O})/\mathbf{R}(\mathbf{X}) - \dim\ \mathbf{R}^{-1}(\mathbf{O})/\mathbf{S}(\mathbf{X}) \,. \tag{5.14.14}$$

In particular we can say that $T \in B(X)^n$ is *Taylor Weyl* if it is Taylor Fredholm with

$$\mathrm{Index}(T) = 0 \,.$$

For "Taylor Browder" tuples we look for the notions of "ascent" and "descent", which have been formulated by Derek Kitson. If $E \subseteq L(X)$ is an arbitrary set of linear operators we shall write, for each $r \in \mathbb{N}$,

$$E^{(r)} = \{T_1 T_2 \ldots T_r : T \in E^r\} \,, \tag{5.14.15}$$

with also

$$E^{(0)} = \{I\} \,.$$

Now we write

$$N^r(E) = N(E^r) = \bigcap\{S^{-1}(0) : S \in E^{(r)}\} \tag{5.14.16}$$

and

$$R^r(E) = R(E^{(r)}) = \sum\{S(X) : S \in E^{(r)}\} \,. \tag{5.14.17}$$

As in the case of a singleton $E = \{T\}$ the null spaces and ranges form an increasing and a decreasing sequence, each of which may or may not stabilise. However the definition of ascent and descent is here a little more subtle: we define

$$\mathrm{ascent}(E) = \inf\{r \in \infty : N^r(E) \cap R^r(E) = \{0\}\} \tag{5.14.18}$$

and

$$\text{descent}(E) = \inf\{r \in \infty : N^r(E) + R^r(E) = X\} .\qquad(5.14.19)$$

This gives back the traditional concept for a singleton $E = \{T\}$; we also still find that if the ascent and the descent are both finite then they are equal. For a tuple

$$T = (T_1, T_2, \ldots, T_n) \in L(X)^n$$

we define the ascent and descent by (5.14.18) and (5.14.19) for the set

$$E = \{T_1, T_2, \ldots, T_n\} \subseteq L(X) ,$$

and the declare $T \in L(X)^n$ to be a *Taylor Browder* tuple if it is Taylor Fredholm with finite ascent and descent.

5.15 Joint Boundary Spectrum

In spite of (5.3.12) writing, for Banach spaces X, Y and Z,

$$BL(X, Y, Z) = \{(S, T) \in BL(Y, Z) \times BL(X, Y) : ST = 0\} ,\qquad(5.15.1)$$

then if (S, T) and (S_n, T_n) in $BL(X, Y, Z)$ satisfy

$$\|S - S_n\| + \|T - T_n\| \to 0 \, (n \to \infty)\qquad(5.15.2)$$

then there is implication

$$S \text{ open} , \ T \text{ boundedbelow} , \ (S_n, T_n) \text{ exact}\qquad(5.15.3)$$

implies

$$(S, T) \text{ exact} .\qquad(5.15.4)$$

This means that if (S, T) is in the topological boundary of the set of exact pairs, open relative to the closed set $BL(X, Y, Z)$, then either S fails to be open or T fails to be bounded below. It follows that, for a commuting pair $T = (T_1, T_2) \in B(X)^2$ of bounded operators, there is inclusion

$$\partial \tau^{Taylor}(T) \subseteq \tau^{left}(T) \cup \iota^{right}(T) .\qquad(5.15.5)$$

We can also show, still with $n = 2$, that

$$\partial \sigma^{Taylor}(T) \subseteq \tau^{Taylor}(T) ,\qquad(5.15.6)$$

and hence also

$$\partial \sigma^{Taylor}(T) \subseteq \partial \tau^{Taylor}(T) \subseteq \sigma^{left}(T) \cup \sigma^{right}(T) . \tag{5.15.7}$$

Curto has an argument here, which relies on index continuity:

$$\text{index}(0, S, T, 0) = \lim_n \text{index}(0, S_n, T_n, 0) = 0 . \tag{5.15.8}$$

Raul Curto also has an example which shows that this boundary result does not extend to tuples with $n \geq 3$. However, in a C*algebra A, we can show that, for commuting tuples $a \in A^n$,

$$\text{iso } \sigma^{Taylor}(a) \subseteq \sigma^{left}(a) \cup \sigma^{right}(a) . \tag{5.15.9}$$

The argument rests on the *Silov idempotent theorem*.

5.16 Bass Stable Rank

We shall say that the ring A has *left stable rank* $\leq n$ if there is implication, for arbitrary $(a, b) \in A^n \times A$,

$$(a, b) \in A_{left}^{-(n+1)} \implies \exists c \in A^n : a - cb \in A_{left}^{-n} , \tag{5.16.1}$$

with a similar definition of "right" stable rank. We observe here that sufficient for this is that the jointly right invertible tuples are dense in A^n with respect to an extension to products of the spectral topology of (3.2.1):

$$A^n \subseteq Cl_{right}^{(n)}(A_{right}^{-n}) , \tag{5.16.2}$$

where

$$Cl_{left}^{(n)}(K) = \{x \in A^n : \forall \text{ finite } J \subseteq A^n \, \exists \, x' \in K \; : 1 - J \cdot (x - x') \subseteq A^{-1}\} , \tag{5.16.3}$$

and

$$Cl_{right}^{(n)}(K) = \{x \in A^n : \forall \text{ finite } J \subseteq A^n \, \exists \, x' \in K \; : 1 - (x - x') \cdot J \subseteq A^{-1}\} , \tag{5.16.4}$$

The argument is rather simple: if, in the notation of (5.1.3), $a' \cdot a + b'b = 1$ with $a' \in Cl_{right}^{(n)}(A_{right}^{-n})$ then there are a'', a''' in A^n with

$$b'b = 1 - a' \cdot a = d - a'' \cdot a \text{ with } d \in A^{-1}, a'' \cdot a''' = 1 ;$$

then $d^{-1}a'' \cdot (a + a'''b'b) = d^{-1}(a'' \cdot a + b'b)$, giving

$$a - cb = a'''d \in A_{left}^{-n} \text{ with } c = -a'''b' .$$

We observe that it is not clear, in contrast to the case $n = 1$, that (5.16.3) and (5.16.4) give the same topology on A^n: Jacobson's lemma (1.8.5) does not extend. Each of these topologies is weaker than the cartesian product of the spectral topologies on each factor A, which coincides with the spectral topology for the direct sum ring A^n.

It is tempting to express the condition (5.16.1) spectrally: if we write

$$\omega^{left}(a, b) = \{(\lambda, \mu) \in \mathbb{C}^{n+1} : \forall c \in A^n : 0 \in \sigma^{left}(a - \lambda + c(b - \mu))\} ,$$
(5.16.5)

then we have inclusion

$$\omega^{left}(a, b) \subseteq \sigma^{left}(a, b) .$$
(5.16.6)

A stronger version of the implication (5.16.1) would require

$$c \in \mathrm{comm}(a, b, c) \cap A^n .$$

If for example the algebra A is commutative then

$$\omega^{left}(a.b) = \bigcap \{\lambda + \nu\mu\} : (\lambda, \mu, \nu) \in \sigma(a, b, c)\} .$$
(5.16.7)

5.17 Determinant and Adjugate

Our definition of "exactness" is not confined to true "complexes", and hence "Taylor invertibility" and the "Taylor spectrum" are meaningful for non-commuting systems of operators and algebra elements. Without commutivity however, the "forward spectral mapping theorem" is liable to fail. The simplest example is again the pair $a = (a_1, a_2) = (e, f) \in A^2$ of 2×2 matrices of (5.1.9); the Taylor spectrum of a is now the non-empty set $\{(0, 0)\}$, and the single polynomial $p = z_1z_2 + z_2z_1 \in \mathrm{Poly}_2$ furnishes the counterexample. Indeed there is implication

$$(0, 0) \notin \sigma^{middle}(e, f) \implies 1 \in (eA + Af) \cap (Ae + fA) ,$$
(5.17.1)

and then, with $p = z_1z_2 + z_2z_1$, there is implication

$$p\sigma^{middle}(e, f) \implies \{p(0, 0)\} = \{0\} \nsubseteq \{1\} = \sigma p(e, f) .$$
(5.17.2)

Here of course, in the sense (1.14.3),

$$(\lambda, \mu) \notin \sigma^{middle}(a, b) \Longleftrightarrow (a - \lambda, b - \mu) \text{ middle invertible} . \tag{5.17.3}$$

For another example we may take $(a, b) = (u, v)$ to be the forward and backward shifts on ℓ_2, or more general ℓ_p.

Specialising the discussion of Sect. 5.13, Taylor invertibility for a pair $(b, c) \in A^2$ of linear algebra elements is in general defined by the (splitting) exactness of its "Koszul complex" $(0, T^\sim, T, 0)$, where arrows run from right to left and

$$T = \begin{pmatrix} b \\ c \end{pmatrix} , \quad T^\sim = \begin{pmatrix} c & -b \end{pmatrix} , \tag{5.17.4}$$

which means, for exactness, or splitting exactness, that the pair (b, c) is left, middle and right, non-singular, or invertible. If we now attack the pair (b, c) with a pair $(p, q) \in \text{Poly}_2^2$ of two-variable polynomials without constant term, then we will replace T and $T^\sim$ with S and $S^\sim$, where

$$S = UT ; \quad S^\sim U = |U|T^\sim ; \quad S^\sim = T^\sim U^\sim ; \quad U^\sim S = T|U| , \tag{5.17.5}$$

with

$$U \in A^{2\times2} ; \quad |U| \in A ; \quad A^\sim \in A^{2\times2} . \tag{5.17.6}$$

If $(p(b, c), q(b, c))$ is Taylor invertible then there will be R and $R^\sim$ for which

$$R^\sim S = 1 = S^\sim R \in A \tag{5.17.7}$$

and

$$RS^\sim + SR^\sim = I = \begin{pmatrix} 1 & 0 \\ 0 & 1 \end{pmatrix} \in A^{2\times2} , \tag{5.17.8}$$

transmitting in particular left and right invertibility to the pair (b, c):

$$(R^\sim U)T = 1 = T^\sim(U^\sim R) \in A , \tag{5.17.9}$$

and then also

$$RT^\sim U^\sim + UTR^\sim = I \in A^{2\times2} . \tag{5.17.10}$$

We claim that (5.17.10) can be converted to

$$U^\sim RT^\sim + TR^\sim U = I \in A^{2\times2} , \tag{5.17.11}$$

making $(T^\sim, T)$ (splitting) exact and hence (b, c) middle invertible. Specifically, if

$$\{w_1, w_2\} \subseteq \text{comm}(b, c) , \tag{5.17.12}$$

then

$$(w_1 \; w_2) \begin{pmatrix} b \\ c \end{pmatrix} = 1 = (c \; -b) \begin{pmatrix} w_2 \\ -w_1 \end{pmatrix} \tag{5.17.13}$$

if and only if

$$\begin{pmatrix} b \\ c \end{pmatrix} (w_1 \; w_2) + \begin{pmatrix} w_2 \\ -w_1 \end{pmatrix} (c \; -b) = \begin{pmatrix} 1 & 0 \\ 0 & 1 \end{pmatrix} . \tag{5.17.14}$$

Indeed if in (5.17.7) R and $R^\sim$ confer on S the commutivity conditions of (5.8.12), then there is two-way implication

$$(5.17.7) \Longleftrightarrow (5.17.8) .$$

If in (5.17.5) we have

$$U = \begin{pmatrix} u_{11} & u_{12} \\ u_{21} & u_{22} \end{pmatrix} \tag{5.17.15}$$

then, fixing U as in (5.17.15) with mutually commuting u_{ij} and varying b, c in comm(u_{ij}), then (5.17.5) is uniquely satisfied by

$$|U| = u_{11}u_{22} - u_{21}u_{12} , \quad U^\sim = \begin{pmatrix} u_{22} & -u_{12} \\ -u_{21} & u_{11} \end{pmatrix} . \tag{5.17.16}$$

This argument offers a derivation, for 2×2 matrices with commuting entries, of the determinant and the adjugate, and demonstrates, for commuting pairs of linear algebra elements, the forward spectral mapping theorem for the Taylor split spectrum.

From (5.17.5) it follows, with

$$M_U = \begin{pmatrix} 1 & 0 & 0 \\ 0 & U & 0 \\ 0 & 0 & |U| \end{pmatrix} , \quad N_T = \begin{pmatrix} 0 & 0 & 0 \\ T & 0 & 0 \\ 0 & T^\sim & 0 \end{pmatrix} , \quad \begin{pmatrix} 0 & R^\sim & 0 \\ 0 & 0 & R \\ 0 & 0 & 0 \end{pmatrix} , \tag{5.17.17}$$

that we have

$$M_U N_T = N_S M_U , \tag{5.17.18}$$

and, by (5.17.7) and (5.17.8),

$$N_{\tilde{R}} M_U N_T + N_T M_U N_{\tilde{R}} = I .\tag{5.17.19}$$

Essentially the same algebra, working with (5.13.17) and (5.10.17), transfers the forward spectral mapping theorem for the Taylor split spectrum from commutative $b \in A^n$ to commutative $(b, c) \in A^{n+1}$.

Enrico Boasso has noticed a certain "quasi-commutative" extension of this: returning to (5.17.4), if also $a = T^\sim T = cb - bc$ and

$$Ta = \Delta_a T , \quad T^\sim \Delta_a = aT^\sim ,\tag{5.17.20}$$

then (arrows running from right to left),

$$\left(0, \left(T^\sim \ a\right), \begin{pmatrix} T & -\Delta_a \\ -1 & T^\sim \end{pmatrix}, \begin{pmatrix} a \\ T \end{pmatrix}, 0\right)\tag{5.17.21}$$

is a complex: each product vanishes. With again $S = UT$, and $d = S^\sim S$, we claim

$$\begin{pmatrix} d \\ S \end{pmatrix} = \begin{pmatrix} |U| & 0 \\ 0 & U \end{pmatrix} \begin{pmatrix} a \\ T \end{pmatrix} ,\tag{5.17.22}$$

$$\begin{pmatrix} S & -\Delta_d \\ -1 & S^\sim \end{pmatrix} \begin{pmatrix} |U| & 0 \\ 0 & U \end{pmatrix} = \begin{pmatrix} UU^\sim U & 0 \\ 0 & U \end{pmatrix} \begin{pmatrix} T & -\Delta_a \\ -1 & T^\sim \end{pmatrix} ,\tag{5.17.23}$$

$$\left(S^\sim \ d\right) \begin{pmatrix} UU^\sim U & 0 \\ 0 & |U| \end{pmatrix} = |U|^2 \left(T^\sim a\right) .\tag{5.17.24}$$

For splitting exactness of the sequence derived from (5.17.22) by replacing T by S we ask

$$\left(\rho \ R^\sim\right) \begin{pmatrix} d \\ S \end{pmatrix} = 1 = \left(S^\sim \ d\right) \begin{pmatrix} R \\ \rho \end{pmatrix} ,\tag{5.17.25}$$

$$\begin{pmatrix} d \\ S \end{pmatrix} \left(\rho \ R^\sim\right) + \begin{pmatrix} R^\sim & 0 \\ 0 & R \end{pmatrix} \begin{pmatrix} S & -\Delta_d \\ -1 & S^\sim \end{pmatrix} = \begin{pmatrix} 1 & 0 & 0 \\ 0 & 1 & 0 \\ 0 & 0 & 1 \end{pmatrix} ,\tag{5.17.26}$$

$$\begin{pmatrix} S & -\Delta_d \\ -1 & S^\sim \end{pmatrix} \begin{pmatrix} R^\sim & 0 \\ 0 & R \end{pmatrix} + \begin{pmatrix} R \\ \rho \end{pmatrix} \left(S^\sim \ d\right) = \begin{pmatrix} 1 & 0 & 0 \\ 0 & 1 & 0 \\ 0 & 0 & 1 \end{pmatrix} .\tag{5.17.27}$$

This now confers splitting exactness on the original sequence (5.17.21).

The factorization (5.17.5) relies heavily on the commutivity of the pair (b, c): if for example

$$U = \begin{pmatrix} c & 0 \\ 0 & b \end{pmatrix} \ , \ S = \begin{pmatrix} cb \\ bc \end{pmatrix} \tag{5.17.28}$$

then equality

$$S^{\sim} U = |U| T^{\sim}$$

with $|U| \in A$ would require

$$\left(bc^2 - cb^2\right) = \left(|U|c - |U|b\right) ,$$

and hence

$$(bc - |U|)c = 0 = (cb - |U|)b \ ;$$

now if also $\{b, c\} \subseteq A^{-1}_{right}$ are each right invertible then it follows

$$cb = |U| = bc . \tag{5.17.29}$$

Thus if there is to be an extended factorization (5.17.5), with $a \neq 0$, then b and c cannot both have a right inverse.

Many Variables

6

In the passage from finite n-tuples $a = (a_1, a_2, \ldots, a_n) \in A^n$ of linear algebra elements to more general systems $a = (a_x)_{x \in X}$, a door opens to the possibility of additional structure on the indexing material X.

6.1 Infinite Systems

If X is a set then a system

$$a = (a_x)_{x \in X} \in A^X$$

can be conceived of as a map

$$a^\wedge : x \mapsto z_x(a) = a_x \quad (X \to A) .$$

Now we have a new possibility: the set X can have structure—topological, linear or semigroup—and then the mapping $a^\wedge$ may or may not respect this, and be either continuous, linear or a homomorphism. Here then is a little surprise: if ω is some kind of "spectrum",

$$a \in A^X \mapsto \omega(a) \subseteq \mathbb{C}^X$$

then, provided ω is subject to the forward spectral mapping theorem (5.1.6),

$$p\omega(a) \subseteq \omega p(a) ,$$

any such structure on X which is respected by a is respected by everything in $\omega(a)$:

$$p \in \text{Poly} ; \ \lambda \in \omega(a) \ : \ p(a) = 0 \Longrightarrow p(\lambda) = 0 . \tag{6.1.1}$$

© The Author(s), under exclusive license to Springer Nature Switzerland AG 2023 151
R. Harte, *Spectral Mapping Theorems*,
https://doi.org/10.1007/978-3-031-13917-8_6

Taking successively

$$p = z_{x'} - z_x \; ; \; p = z_{x''} - \alpha_x z_x - \alpha_{x'} z_{x'} \; ; \; p = z_{x''} - z_x z_{x'} \; ; \; p = 1 \qquad (6.1.2)$$

gives continuity, linearity, the multiplicative property and respect for the identity.

As a fundamental example of this phenomenon take the set X to be the algebra A itself, and take $a \in A^X$ to be the system for which $a^\wedge = I : A \to A$ is the identity mapping:

$$a_x = x \; (x \in X = A) . \qquad (6.1.3)$$

Evidently $a^\wedge$ is continuous, linear and multiplicative, and hence

$$\omega(a) \subseteq \sigma(A) \qquad (6.1.4)$$

is a subset of the "maximal ideal space" $\sigma(A)$ of (5.4.3). It follows that Gelfand's theorem for commutative Banach algebras is indeed a special case of the spectral mapping theorem.

The spectral mapping theorem, and hence Gelfand's theorem, evidently extend to Banach algebras A which are "quasicommutative" in the sense that

$$[A, [A, A]] = \{0\} , \qquad (6.1.5)$$

where we write

$$[a, b] = ab - ba .$$

We are of course extending the quasicommutivity (5.8.7) from n tuples $a \in A^n$ to arbitrary systems $a \in A^X$, in particular with $X = A$. Explicitly, for arbitrary $(a, b, c) \in A^3$,

$$(ab - ba)c = c(ab - ba) .$$

The "non commutative polynomials" of Sect. 5.1 now extend to "polynomials in X variables" Poly_X, and systems $p \in \text{Poly}_X^Y$ of such polynomials induce mappings

$$a \mapsto p(a) : A^X \to A^Y .$$

If, generalizing (6.1.3), we take $X \subseteq A$ to be a subset, then we shall write

$$\omega(a) = \omega(X) , \qquad (6.1.6)$$

in particular for $\omega = \sigma^{left}$. Thus if $\lambda \in \mathbb{C}^X$ is arbitrary then there is implication

$$A \neq N_\lambda(X) \Longrightarrow \lambda \in \sigma^{left}(a) \Longrightarrow 1 \notin N_\lambda(X) , \qquad (6.1.7)$$

where we write

$$N_\lambda(X) = \sum_{x \in X} A(x - \lambda_x) ; \qquad (6.1.8)$$

in particular $N_0(X)$ is the left ideal generated by X. We now observe an inclusion: for arbitrary $\lambda \in \mathbb{C}^X$

$$[X, X] \subseteq N_\lambda(X) \text{ and } X \subseteq N_\lambda(X)^{-1} N_\lambda(X) , \qquad (6.1.9)$$

and hence

$$[X, X]\mathrm{alg}(X) \subseteq N_\lambda(X) . \qquad (6.1.10)$$

For (6.1.10) we argue that, for arbitrary $\{x, y\} \subseteq X$

$$[x, y] \equiv xy - yx = [x - \lambda_x, y - \lambda_y] \in N_\lambda(X) ,$$

while also

$$(x - \lambda_x)y = (x - \lambda_x)(y - \lambda_y) + \lambda_y(x - \lambda_x) \in N_\lambda(X) .$$

It follows that there is implication

$$\sigma^{left}(X) \neq \emptyset \implies N_0([X, X]\mathrm{alg}(X)) \neq A . \qquad (6.1.11)$$

For example, replacing the set X by the left ideal $N_0(X)$, there is equivalence

$$0 \in \sigma^{left} N_0(X) \iff 1 \notin N_0(X) \iff 0 \in \sigma^{left}(X) . \qquad (6.1.12)$$

6.2 Vector Valued Spectra

If X and Y are linear vector spaces *homogeneous polynomials* p *of degree* n from X to Y are derived from *multilinear mappings*

$$P : X^n \to Y \qquad (6.2.1)$$

by setting

$$p(x) = P(x, x, \ldots, x) . \qquad (6.2.2)$$

Conversely we can recover a *symmetric version* $P = p^\wedge$ from p by *polarization*. In turn the multilinear $P : X^n \to Y$ can be expressed as a *linear* mapping, to the space

Y, from the tensor product of n copies of X. When X and Y are Banach spaces then bounded multilinear mappings generate bounded polynomials:

$$\|p\| \leq \|p^\wedge\| \leq \frac{n^n}{n!} \|p\| \ . \tag{6.2.3}$$

Suppose that A is a Banach algebra and that E is a Banach space: then bounded linear functionals $f \in E^*$ act on a uniform cross-normed product $A \otimes E$, by continuous linear extension of the mapping

$$a = \sum_{j \in J} a_j \otimes x_j \mapsto f^\vee(a) = \sum_{j \in J} f(x_j)a_j \in A \ . \tag{6.2.4}$$

Turning this inside out, elements $a \in A \otimes E$ induce mappings

$$a^\wedge : f \mapsto f^\vee(a) \in A \ . \tag{6.2.5}$$

In other words elements $a \in A \otimes E$ can be regarded as systems of elements of the algebra A, indexed by the dual space E^*. Evidently $a^\wedge$ is bounded and linear: it follows, going back to (6.1.2), that

$$\omega(a^\wedge) \subseteq E^{**} \tag{6.2.6}$$

is now a subset of the second dual of the Banach space E. If we look more closely at this, we can observe that the mapping $a^\wedge$ is actually *weak* continuous*, from E^* into the Banach algebra A, which therefore applies to elements of $\omega(a^\wedge)$, which therefore (3.6.14) lie in the image of E in E^{**}:

$$\omega(a) \subseteq E \ . \tag{6.2.7}$$

In words, the left, and the right, spectrum of a tensor product element $a \in A \otimes E$ is now a compact subset of the Banach space E. Formally, if $a \in A \otimes E$,

$$\sigma^{left}(a) = \{x \in E : 1 \notin \sum_{f \in E^*} A(f^\vee(a) - f(x))\} \ . \tag{6.2.8}$$

For example,

$$\sigma_A^{left}(1 \otimes x) = \{x\} \ .$$

Now the "polynomials" $p : A^X \to A$ we have looked at in Sect. 6.1 were indeed polynomials in "infinitely many variables", although in practice only acting on finitely many of them. By contrast if $X = E$ is infinite dimensional the polynomials of (6.2.1) are truly infinite: for example linear functionals on E count as polynomials which are homogeneous of degree one. We now claim that bounded polynomials $p : E \to F$ can be extended to mappings

$$I \otimes p : A \otimes_1 E \to A \otimes_1 F . \tag{6.2.9}$$

We have also, for each $\omega \in \{\sigma^{left}, \sigma^{right}\}$, the one-way inclusion (5.1.6):

$$p \, \omega(a) \subseteq \omega \, p(a) \subseteq F . \tag{6.2.10}$$

When in particular the tensor product element $a \in A \otimes_1 E$ is *commutative*, in the sense that

$$\{f^\vee(a) : f \in E^*\} \subseteq A \tag{6.2.11}$$

is commutative, then there is equality in (6.2.10). Indeed (6.2.10) follows from the remainder theorem for vector polynomials, while equality for commutative tensor product elements comes from the same kind of argument as for (5.3.8).

If, in particular, $E = B$ is also a Banach algebra, so that $D = A \otimes E = A \otimes B$ too is an algebra, then there are three spectrums

$$\omega_D(a) \subseteq \mathbb{C} \; ; \; \omega_A(a) \subseteq B \; ; \; \omega_B(a) \subseteq A .$$

These are of course interrelated: if $a \in D$ is commutative then, by (5.3.10) together with (5.2.3),

$$\omega_D(a) = \bigcup \{b \in \omega_A(a) : \omega_B(b)\} . \tag{6.2.12}$$

In the terminology of Graham Allan, characters $f \in \sigma(A)$ induce "$1 \otimes B$ characters" $f^\vee : A \otimes B \to B$ on $A \otimes B$; if in particular the algebra A is commutative then the product $A \otimes B$ is "$1 \otimes B$ commutative" in $A \otimes B$.

More abstractly, if $1 \in D \subseteq A$, we can attempt to define, for $a \in A$, a "D-spectrum"

$$\sigma(a, D) \equiv \sigma_A(a, D) \subseteq D \tag{6.2.13}$$

for which

$$a \in D \Longrightarrow \sigma(a, D) = \{a\} , \tag{6.2.14}$$

subject to the forward and backward spectral mapping theorems for polynomials, with for homomorphisms $T : A \to B$ and subalgebras $T(D) \subseteq E \subseteq B$,

$$T^{-1} \sigma_B(Ta, E) \subseteq \sigma_A(a, D) . \tag{6.2.15}$$

We might also expect, when $1 \in B \subseteq D \subseteq A$,

$$\sigma_A(a, B) = \bigcup \{\sigma_D(d, B) : d \in \sigma_A(a, D)\} . \tag{6.2.16}$$

One rather obvious candidate would be to take, for (6.2.13),

$$\sigma_A(a, D) = \{d \in D : a - d \notin A^{-1}\} . \tag{6.2.17}$$

This rather quickly fails ($a = 0 \in D = A$) the test (6.2.14), and also the spectral mapping theorem for polynomials: for the forward inclusion set

$$a = \begin{pmatrix} 0 & 1 \\ 0 & 0 \end{pmatrix} , \quad d = \begin{pmatrix} 0 & 0 \\ 1 & 0 \end{pmatrix} , \quad p = z^2 ;$$

while for the backward

$$a = \begin{pmatrix} 0 & 1 \\ 1 & 0 \end{pmatrix} , \quad d = \begin{pmatrix} 0 & 0 \\ 1 & 0 \end{pmatrix} , \quad p = z^2 .$$

If the D-spectrum generated by (6.2.17) is liable to be too big, then the candidate

$$\{\Phi(a) : \Phi \in \sigma(a, D)\} \tag{6.2.18}$$

given by the D-characters of Graham Allen,

$$\sigma(A, D) = \{\Phi \in HBL(A, D) : D \subseteq (I - \Phi)^{-1}(0)\} , \tag{6.2.19}$$

is liable to be too small. As a compromise we return to the relative and restricted spectra of Sect. 5.2 and offer the definition

$$\sigma^{left}(a, D) = \{d \in D : \sigma^{left}(d) = \sigma^{left}_{a-d=0}(d)\} , \tag{6.2.20}$$

with the corresponding recipe for the right D-spectrum. For example if $A = D = C(\Omega)$ then necessary and sufficient for $d \in \sigma^{left}(a., D)$ is that

$$\forall t \in \Omega , \ \exists s \in \Omega : d(t) = d(s) = a(s) : \tag{6.2.21}$$

the graphs of a and d intersect at least once. With this definition the D-spectrum is "commutatively closed" and subject to the forward spectral mapping theorem;

$$\sigma^{left}(a, D) = \mathrm{cl}_{comm} \sigma^{left}(a, D) ,$$

where, if $K \subseteq A$,

$$\mathrm{cl}_{comm}(K)) = \{a \in A : a \in \mathrm{cl}(K \cap \mathrm{comm}(a)\} .$$

There is inclusion

$$\{\Phi(a) : \Phi \in \sigma(A, D)\} \subseteq \sigma^{left}(a, D) \subseteq \{d \in D : 0 \in \sigma^{left}(a - d)\} . \quad (6.2.22)$$

If $D \subseteq A$ is an Allen subalgebra then

$$\sigma^{left}(a, D) = \bigcup_{\Phi \in \sigma(A, D)} \sigma_D^{left}(\Phi(a), D) = \bigcup_{\varphi \in \sigma(D)} \sigma_D^{left}(\varphi^\wedge(a), D) , \quad (6.2.23)$$

and if $a \in D$ then, rather than (6.2.1),

$$a + D \cap \mathrm{Rad}(A) \subseteq \sigma_A^{left}(a, D) . \quad (6.2.24)$$

To see the commutative closedness (6.2.21) argue that if $d = \lim_n d_n$ with $d_n d = dd_n$ and $\lambda \in \sigma^{left}(d)$ then there are λ_n with

$$(\lambda, \lambda_n) \in \sigma^{left}(d, d_n) , \implies \lambda - \lambda_n \in \sigma^{left}(d - d_n) \text{ and } \lambda_n \in \sigma^{left}(d_n) ,$$

giving back an old theorem of Newburgh. Now if $d_n \in \sigma(a, D)$ then $(\lambda_n, \lambda_n) \in \sigma^{left}(d_n, a)$, and then by upper semi-continuity $(\lambda, \lambda) \in \sigma^{left}(d, a)$. Now if $d \in \sigma^{left}(a, D)$ and $\mu \in D$ is in $\sigma^{left} p(d)$ then we can write $\mu = p(\lambda)$ with $\lambda \in \sigma^{left}(d)$ and hence $(\lambda, \lambda) \in \sigma^{left}(a, d)$, giving $(\mu, \mu) \in \sigma^{left}(f(a), f(d))$. Next if $(\lambda, \lambda) \in \sigma^{left}(a, d)$ then $0 = \lambda - \lambda \in \sigma^{left}(a, d)$. Also if $\Phi(a)$ were not in $\sigma^{left}(a, D)$ then there would for $\lambda \in \sigma^{left} \Phi(a)$ be a' and d' in A for which $a'(a - \lambda) + d'(\Phi(a) - \lambda) = 1$, giving $\Phi(a' + d')(\Phi(a) - \lambda) = 1$. Now apply Allen's theorem with (a, d) in place of a. Finally if $a \in D$ and $d \in D$ with $a - d \in \mathrm{Rad}(A)$ then $\lambda \in \sigma^{left}(d)$ means that $d - \lambda \in N$ for some maximal left ideal $N \in ML(A)$; since $\mathrm{Rad}(A) = \bigcap ML(A)$ it follows $a - d \in N$, so that $a - \lambda$ and $d - \lambda$ are both in N. It is not immediately clear how to extend the definition (6.2.17) of $\sigma(a, D)$ to arbitrary systems $a \in A^X$; as it stands it would mean that if the spectrum of $d \in D^X$ is empty then d is automatically in the D-spectrum of $a \in A^X$. It would have been pleasant to find that the D spectrum of the system $A = (a)_{a \in A}$ coincided with the D characters $\sigma(A, D)$.

6.3 Waelbroeck Algebras

A *Waelbroeck algebra* is a topological algebra for which the invertible group A^{-1} is an open set, with the inversion map $z^{-1} : a \mapsto a^{-1} (A^{-1} \to A)$ continuous. Antoni Wawrzynczyk has deployed Schur's Lemma to extend the spectral mapping theorem (5.3.11) to locally convex Waelbroeck algebras. Much of Banach algebra theory persists in Waelbroeck algebras; however the argument of §5.3 is not available: the topological boundary ∂A^{-1} of the invertible group need no longer lie among the topological zero divisors, and we must therefore go back to the consequences of Liouville's theorem (4.3.4). The reader may possibly wonder why, in its "spectral topology", as given by (3.2.1), a complex linear algebra does not also qualify as a Waelbroeck algebra; but for the continuity of scalar mutiplication in this case we have to also impose the spectral topology on the scalar field.

The Wawrzynczyk argument consists in showing that the implication (6.1.11) goes both ways.

Thus suppose that A is a (complex) locally convex Waelbroeck algebra and that $X \subseteq A$, and write $\mathbf{J}_X$ for the set those proper left ideals $1 \notin J = N_0(J) \subseteq A$ for which

$$[X, X] \subseteq J \text{ and } X \subseteq J^{-1}J : \tag{6.3.1}$$

we claim that if $J \in \mathbf{J}_X$ there is a maximal left ideal $M \in ML(A) \cap \mathbf{J}_X$ and $\lambda \in \sigma^{left}(X)$ for which

$$J + N_\lambda(X) \subseteq M . \tag{6.3.2}$$

By Zorn's Lemma $\mathbf{J}_X$ has a maximal element M, and we shall show that also $M \in ML(A)$ is maximal among proper left ideals of A. Now if $a \in A$ and $x \in X$ define

$$[L_a R_x]_M : [b]_M \mapsto [abx]_M \ (A/M \to A/M) . \tag{6.3.3}$$

We claim that the only subspaces $V \subseteq A/M$ invariant under

$$[L_A R_X]_M = \{[L_a R_x]_M : (a, x) \in A \times X\}$$

are $O = \{0\}$ and A/M. To see this note that, if $\pi_M : A \to A/M$ is the quotient map, then

$$J = \pi_M^{-1}(V) \in \mathbf{J}_X .$$

By maximality $J = M$ or $J = A$. By Schur's Lemma (1.7.15), the commutant

$$D = \text{comm}[L_A R_X]_M \subseteq L(A/M) , \tag{6.3.4}$$

among all linear maps on A/M, is a division algebra. We can represent D in terms of elements of A: with

$$E = \{a \in M^{-1}M : [a, X] \subseteq M\} , \tag{6.3.5}$$

we find that $1 \in E \subseteq A$ is a subalgebra, with $M \subseteq E$ a (proper) two-sided ideal, and now

$$D = [R_E]_M = \{[R_a]_M : a \in E\} . \tag{6.3.6}$$

Since $E \subseteq A$ is inverse closed then both E and $D = E/M$ are also Waelbroeck algebras. The Gelfand-Mazur lemma for locally convex Waelbroeck algebras gives $D \cong \mathbb{C}$ so that R_x is scalar, for each $x \in X$, and hence $[L_A]_M = [L_A R_X]_M$ is also irreducible on A/M, giving indeed $M \in ML(A)$: thus finally

$$N_0([X, X]\mathrm{alg}(X)) \neq A \Longrightarrow \sigma^{left}(X) \neq \emptyset , \tag{6.3.7}$$

reversing the implication (6.1.11). The "projection property" for σ^{left} now follows: if

$$X \subseteq A , \ \lambda \in \sigma^{left}(X) \ and \ y \in \mathrm{comm}(X) \subseteq X^{-1}X$$

then

$$1 \notin J = N_\lambda(X) \in \mathbf{J}_X$$

and by (6.3.2) there is $M \in ML(A)$ and $\mu \in \mathbb{C}$ for which

$$N_\lambda(X) + N_\mu(\{y\}) \subseteq M \subseteq A ,$$

giving

$$(\lambda, \mu) \in \sigma^{left}(X \cup \{y\}) . \tag{6.3.8}$$

The argument for the right spectrum follows at once, by "reversal of products", from the argument for the left; alternatively reinterpret the left ideals $N_\lambda(X)$ as right ideals, and also reverse the residual quotient in (6.3.1). For more abstract X, and $a \in A^X$, apply the argument above to $\{a_x : x \in X\} \subseteq A$.

Towards the corresponding discussion of the split Taylor spectral mapping theorem, we should instead replace left ideals $N_\lambda(a)$ by sums of left and right ideals, as in for example (1.7.12), and then pass from $a \in A^X$ to a transfinite version of Λ_a, as in (5.13.17): we wish the reader luck.

6.4   Categories and Functors

If A is a complex Banach algebra with identity then systems

$$a = (a_x)_{x \in X} \tag{6.4.1}$$

can be regarded as the *objects* of the *category*

$$A^{\text{SET}} \tag{6.4.2}$$

of all mappings from sets to the Banach algebra A. The *morphisms* of this category are, in the first instance, everything that can be induced by mappings $j : Y \to X$, sending

$$a \mapsto a \circ j \; (A^X \to A^Y) \,. \tag{6.4.3}$$

However there is a richer category A^{SET} with the same objects, whose morphisms are *systems of polynomials*

$$p = (p_y)_{y \in Y} \in \text{Poly}_X^Y \,; \tag{6.4.4}$$

the mapping $j : Y \to X$ is represented by the system

$$(z_{jy})_{y \in Y} \tag{6.4.5}$$

of coordinate polynomials. Subcategories of A^{SET} include the *finite* systems **FIN** A^{SET} and the *commuting* systems **COM** A^{SET}, and their intersection the finite commuting systems. Specialising to the Banach algebra $A = \mathbb{C}$, there is also a category **SUB** $\mathbb{C}^{\text{SET}}$ whose objects are *sets* of systems of complex numbers, with polynomial morphisms; subcategories include sets of finite systems, and *compact* sets of systems. With this background we suggest that a *joint spectrum* should be a *functor*

$$\omega : \textbf{COM } A^{\text{SET}} \to \textbf{SUB } \mathbb{C}^{\text{SET}} \,. \tag{6.4.6}$$

In words commuting systems $a \in A^X$ are mapped into subsets $\omega(a) \subseteq \mathbb{C}^X$, while systems of polynomials $p : A^X \to A^Y$ generate

$$\omega(p) : \omega(a) \mapsto \omega p(a) \,. \tag{6.4.7}$$

In a sense the functorial property of ω resides in the spectral mapping theorem.

Suppose now that the set $X = F(K)$ is the image, under a *forgetful functor*

$$F : \Omega \to \textbf{SET} \,, \tag{6.4.8}$$

of some object K from another category Ω, and suppose that there is another "forgetful functor"

$$G : \textbf{BALG} \to \Omega \tag{6.4.9}$$

from the category of Banach algebra homomorphisms; for example Ω could be linear mappings between vector spaces, continuous mappings between topological spaces, or semigroup homomorphisms. What our surprising observation (6.1.1) is telling us is that, in these important special cases, there is implication

$$G \circ a \circ F \in \Omega \Longrightarrow G \circ \omega(a) \circ F \subseteq \Omega . \tag{6.4.10}$$

In words, if $a \in A^X$ can be interpreted as a morphism of the category Ω, then so can everything in its spectrum $\omega(a)$.

The idea of Müller regularity persists here: we shall write, for $a, b \in A^X$,

$$(a \cdot b)_x = a_x b_x$$

and

$$c \in \mathbf{FIN}(A^X) \Longrightarrow \sum(c) = \sum_{x \in X} c_x .$$

For a system of subsets (R_X) in $\mathbf{COM}\ A^X$ to qualify as a "joint regularity" we ask that, for arbitrary $(a, b) \in \mathbf{COM}(A^{X,Y})$,

$$Y = X, a \cdot b \in \mathbf{FIN}(A^X), \sum(a \cdot b) = 1 \Longrightarrow a \in R_X ; \tag{6.4.11}$$

$$a \in R_X \Longrightarrow (a, b) \in R_{X,Y} ; \tag{6.4.12}$$

$$(a, b - \mathbb{C}^Y) \subseteq R_{X,Y} \Longrightarrow a \in R_X . \tag{6.4.13}$$

Joint Müller regularity induces joint spectrum:

$$\varpi_R : a \in \mathbf{COM}(A^X) \mapsto \{\lambda \in \mathbb{C}^X : a - \lambda \notin R_X\} ; \tag{6.4.14}$$

conversely a joint spectrum induces a joint regularity:

$$H_X^\omega = \{a \in \mathbf{COM}(A^X) : 0 \notin \omega(a)\} . \tag{6.4.15}$$

6.5   Functional Calculus

Mathematicians have for many years been trying to capture and tame the intuition of the engineer Oliver Heaviside, which interprets the expression $f(a)$ for functions f and algebra elements a. For arbitrary linear algebra elements $a \in A$ and polynomials $f = p$ the interpretation of $f(a)$ is obvious, and there is an extension to rational functions in which $f(a)$ is interpreted, for the reciprocal $f = z^{-1}$, as the two-sided inverse a^{-1}, provided $a \in A^{-1}$ has such a thing: already here, "spectral disjointness" is required between a and f. For "non-commutative polynomials" $f = p$ in n variables then again the interpretation of $f(a)$ is clear for arbitrary $a \in A^n$. For Banach algebras there was intense activity constructing $f(a)$ whenever $a \in A^n$ is commutative and f is holomorphic on some open neighbourhood of an appropriate spectrum $\sigma(a) \subseteq \mathbb{C}^n$. This, with some effort, was first achieved, by Richard Arens and A.P. Calderon, for commutative Banach algebras. Lucien Waelbroeck, and later Sean Dineen, worked on many-variable extensions of this, involving Banach space tensor products. For non-commutative algebras the Joseph Taylor version τ^{taylor} of the joint spectrum proved able to support a holomorphic calculus for commuting systems of bounded operators on Banach spaces; Florian Vasilescu demonstrated a simplified version of the construction valid for Hilbert space operators. The Vasilescu construction of $f(a)$ extends to Banach space operator systems $a \in A^n = B(X)^n$ when the function f is holomorphic on neighbourhoods of the Taylor split spectrum $\sigma^{taylor}(a)$, and this carries over to arbitrary Banach algebras. At its core the construction is based on Stokes' theorem; for one variable the idea is latent in the Cauchy integral formula and Green's theorem. The definitive account of the Taylor functional calculus, for both the Taylor spectrum of operators, and the Taylor split spectrum, is the book [259] of Vladimir Müller.

6.6   Number Theory

J.E. Littlewood, in his "Miscellany", quotes a nameless savant who maintained that, every once in a while, a scientist should "perform a damfool experiment", such as "playing the trumpet to his tulips". In that spirit, we observe that elementary number theory can be described in language very like spectral theory. With

$$\mathbb{N} = \{1, 2, 3, \ldots\} = \bigcup_{n=1}^{\infty} \mathbb{N}_n \tag{6.6.1}$$

the natural numbers, we have

$$\mathbb{N}_n = \{1, 2, \ldots n\}, \tag{6.6.2}$$

an *initial segment*. Now declare $m \in \mathbb{N}$ to be a *factor* of n, or "divisor", provided

$$n \in \mathbb{N}m. \tag{6.6.3}$$

Equivalently (Green's relation)

$$\mathbb{N}n \subseteq \mathbb{N}m. \tag{6.6.4}$$

The traditional notation is $m|n$; we shall prefer

$$m \in \mathbb{N}_{-1}\{n\} \,. \tag{6.6.5}$$

Here, in contrast to the residual quotients (1.1.8) we write, for $K, H \subseteq A$

$$K_{-1}H = \{x \in A : H \subseteq Kx\} \,; \quad HK_{-1} = \{x \in A : H \subseteq xK\} \,. \tag{6.6.6}$$

It is curious how early in the discussion of the natural numbers, the subtleties of factorization present themselves: as "Uncle Petros" tells his nephew, "multiplication is artificial". The subset $\mathbb{P} \subseteq \mathbb{N}$ of *primes* is fundamental:

$$\mathbb{P} = \{p \in \mathbb{N} : \mathbb{N}_{-1}\{p\} = \{1, p\}\} \setminus \{1\} \,. \tag{6.6.7}$$

The tactical decision to exclude the number 1 from $\mathbb{P}$ is in contrast to our attitude to the empty set $\emptyset$ and the zero subspace $\{0\}$. We shall write, for $n \in \mathbb{N}$,

$$\mathbb{P}_n = \mathbb{P} \cap \mathbb{N}_n \,. \tag{6.6.8}$$

It is a well known, and easily proved, consequence of the "fundamental theorem of arithmetic", that $\mathbb{P}$ is infinite: for if, to the contrary

$$\mathbb{P} \subseteq \mathbb{N}_n \,,$$

then nowhere in the product $n! + 1$ could there be any primes. Thus

$$\mathbb{P} = \{p_1, p_2, p_3 \ldots\} = \{2, 3, 5, 7, \ldots\} \subseteq \mathbb{N} \,, \tag{6.6.9}$$

where, as sequences rather than sets,

$$\mathbf{p} = (p_1, p_2, p_3, \ldots) = (2, 3, 5, 7 \ldots) \in \mathbb{N}^{\mathbb{N}} \,. \tag{6.6.10}$$

There is of course no simple formula for $p_n \in \mathbb{N}$; recursively (*Sieve of Eratosthenes*)

$$p_{n+1} = \text{Min}(\mathbb{N} \setminus \{1\} \setminus p_n \mathbb{N}) \,. \tag{6.6.11}$$

If we reflect that the factorial function

$$n \mapsto n! = 1 \cdot 2 \cdot \ldots \cdot n$$

has a significant extension (the Gamma function) to the complex plane $\mathbb{C}$, then we might wonder if the something similar might be available to the inscrutable $\mathbf{p}$ of (6.6.10). It is now the *fundamental theorem of arithmetic*

$$\mathbb{N} = \prod \mathbb{P} \,, \tag{6.6.12}$$

that every natural number is a (finite) product of primes. If we write, for $n \in \mathbb{N}$ and $p \in \mathbb{P}$ the p-adic valuation of n,

$$v_p(n) = \text{Max}\{k \in \mathbb{N} : p^k \in \mathbb{N}_{-1}\{n\}\} , \qquad (6.6.13)$$

then there is, for each $n \in \mathbb{N}$, $k \in \mathbb{N}$ for which

$$n = p_1^{v_1(n)} p_2^{v_2(n)} \cdots p_k^{v_k(n)} , \qquad (6.6.14)$$

where $v_j(n) = v_q(n)$ with $q = p_j$.

It is sometimes difficult to be sure that $n \in \mathbb{N}$ is prime: but if we can find $p \in \mathbb{P}$ for which

$$p < n < p^2 , \qquad (6.6.15)$$

then we need only search $\mathbb{P}_p$ for factors of n; if there are none then $n \in \mathbb{P}$. This now gives us, in a Littlewood "damfool experiment", a sort of "spectrum" if we set, for $n \in \mathbb{N}$,

$$\varpi(n) = \{p \in \mathbb{P} : p \in \mathbb{N}_{-1}\{n\}\} . \qquad (6.6.16)$$

Evidently

$$\varpi(n) \subseteq \mathbb{P}_n \subseteq \mathbb{N}_n . \qquad (6.6.17)$$

There is two way implication

$$n = 1 \Longleftrightarrow \varpi(n) = \emptyset . \qquad (6.6.18)$$

If $n \in \mathbb{N}$ and $p \in \mathbb{P}$ then there is implication

$$p < n < p^2 \Longrightarrow \varpi(n) \subseteq \mathbb{P}_p \cup \{n\} . \qquad (6.6.19)$$

$n \in \mathbb{N}$ is a *prime power* provided its spectrum is a singleton

$$\#\varpi(n) = 1 , \qquad (6.6.20)$$

and *square free* provided every point of its spectrum has multiplicity one

$$p \in \varpi(n) \Longrightarrow v_p(n) = 1 ; \qquad (6.6.21)$$

thus a square free prime power is itself a prime. The spectral mapping theorem here is "Euclid's lemma", another sort of logarithmic law:

$$\{m, n\} \subseteq \mathbb{N} \Longrightarrow \varpi(mn) = \varpi(m) \cup \varpi(n) . \qquad (6.6.22)$$

Fermat's (little) theorem says that

$$(1 < n \in \mathbb{N} \text{ and } p \in \mathbb{P}) \Longrightarrow p \in \varpi(n) \cup \varpi(n^{p-1} - 1) ; \qquad (6.6.23)$$

Wilson's theorem says that

$$p \in \mathbb{P} \Longrightarrow p \in \varpi(1 + (p - 1)!) . \qquad (6.6.24)$$

Finally, the *Euclidean Algorithm* demonstrates implication

$$\varpi(m) \cap \varpi(n) = \emptyset \Longrightarrow 1 \in \mathbb{Z}m + n\mathbb{Z} : \qquad (6.6.25)$$

once again, spectral disjointness appears to imply exactness. As in linear algebra, our "spectrum" $\varpi(n)$ provides only limited information about n, and more can be gained by observing the "multiplicity" v: indeed the fundamental theorem of arithmetic says that the spectrum, together with multiplicity, completely determines $n \in \mathbb{N}$.

Intermediate between the natural numbers and the primes, the square-free numbers (6.6.21) form a sequence

$$\mathbf{q} = (1, 2, 3, 2, 5, 6, 7, 2, 3, 10, 11, 6, 13, 14, 15, 2, 17, \ldots) \qquad (6.6.26)$$

in which, for each $n \in \mathbb{N}$, q_n is a sort of "square-free root" of n: if $n \in \mathbb{N}$ and $p \in \mathbb{P}$ then

$$\varpi(q_n) = \varpi(n) \text{ and } v_{q_n}(p) \leq 1 . \qquad (6.6.27)$$

Thus q_n records the spectrum of n, discarding multiplicity: indeed we can regard $q_n \in \mathbb{N}$ of as a sort of "vector-valued spectrum" of $n \in \mathbb{N}$ running in parallel with $\varpi(n) \subseteq \mathbb{P}$. At the risk of over-egging the pudding, we could even think of the product

$$n = q'_n q_n \qquad (6.6.28)$$

as a sort of "polar decomposition" of $n \in \mathbb{N}$, and refer to $q'_n = n/q_n$ as the "carapace" of $n \in \mathbb{N}$. Alternatively, transferring some of the material in the square free root to the carapace, we also have

$$n = r'_n r_n , \qquad (6.6.29)$$

in which r_n is square free, while r'_n is a perfect square —specifically the largest square among the factors of n; indeed Rosenthal-cubed have exploited this decomposition to show that the prime reciprocals form a divergent series:

$$\sum_{p \in \mathbb{P}} 1/p = \sum_{n=1}^{\infty} 1/p_n = \infty . \qquad (6.6.30)$$

As a sort of inverse to $\mathbf{p} \in \mathbb{N}^{\mathbb{N}}$, we offer

$$\mathbf{s} = (s_1, s_2, s_3 \ldots) ,$$

where, for each $n \in \mathbb{N}$,

$$s_1 = 1 ; \quad s(p_n) = n + 1 ,$$

and then s_n is derived from the product formula (6.6.14): thus

$$\mathbf{s} = (1, 2, 3, 4, 4, 6, 5, 8, 9, 8, 9, 12, 11, 10, 12, 16, 15, 18, 16 \ldots) .$$

Our spectrum lies in the complement of the "totatives" of n: with

$$\text{Tot}(n) = \{k \in \mathbb{N}_n : k_\wedge n = 1\} , \qquad (6.6.31)$$

where $m_\wedge n$ is the "highest common factor" of m and n:

$$\mathbb{N}_{-1}\{m_\wedge n\} = \mathbb{N}_{-1}\{m\} \cap \mathbb{N}_{-1}\{n\} , \qquad (6.6.32)$$

and we have

$$\varpi(n) = \mathbb{P}_n \setminus \text{Tot}(n) ; \qquad (6.6.33)$$

now *Euler's totient function* ϕ is defined by the formula

$$\phi(n) = \#\text{Tot}(n) . \qquad (6.6.34)$$

For example if $\{p, q\} \subseteq \mathbb{P}$ are distinct primes then

$$\phi(pq) = (p - 1)(q - 1) . \qquad (6.6.35)$$

Complex polynomials in one variable have arithmetic similar to the integers: if

$$p = z^k + \ldots + \alpha_1 z + \alpha_0 \in \text{Poly}_1 \qquad (6.6.36)$$

is a "monic" polynomial, then the fundamental theorem of algebra (2.7.20) says that

$$p \equiv p(z) = \prod_{j=1}^{k} (z - \lambda_j) = \prod_{\lambda \in \mathbb{C}} (z - \lambda)^{\nu_p(\lambda)} : \qquad (6.6.37)$$

here there are possible repetitions among the $\{\lambda_j : j \in \{1, 2, \ldots, k\}\}$, while all but finitely many of the $\nu_p(\lambda)$ vanish:

$$p \in \text{Poly}_1 \subseteq \mathbb{C}[z] \implies \#\{\lambda \in \mathbb{C} : \nu_p(\lambda) \neq 0\} < \infty . \qquad (6.6.38)$$

The "primes" among the monic polynomials are $\{z - \lambda : \lambda \in \mathbb{C}\}$, and $p \in \text{Poly}_1$ has both a "vector-valued" spectrum

$$\{z - \lambda_j : j \in \{1, 2, \ldots k\}\} = \{z - \lambda : \nu_p(\lambda) \neq 0\} , \qquad (6.6.39)$$

and a numerical spectrum

$$\sigma(1/p) = p^{-1}(0) \subseteq \mathbb{C} . \qquad (6.6.40)$$

Determination of the spectrum $\varpi(p)$ of a polynomial p is notoriously difficult if its degree $k \geq 5$ in (6.6.33), but necessary and sufficient for p to be "square free" is that its spectrum be disjoint from that of its derivative dp/dz:

$$\varpi(p) \cap \varpi(dp/dz) = \emptyset . \qquad (6.6.41)$$

More generally, if we write

$$p \equiv p(z) = p^\vee p^\wedge , \qquad (6.6.42)$$

with "square free root" $p^\wedge$ and "carapace" $p^\vee$, then we can show

$$p^\vee = \text{hcf}(p, dp/dz) . \qquad (6.6.43)$$

The Euclidean algorithm (4.10.10) continues to apply: if $\{q, r\} \subseteq \text{Poly}_1$ then

$$q^{-1}(0) \cap r^{-1}(0) = \emptyset \implies 1 \in \mathbb{C}[z]q + r\mathbb{C}[z] . \qquad (6.6.44)$$

The Euclidean algorithm here has an application to the "diagonalization" of a matrix $T \in \mathbb{C}^{k \times k}$: if

$$p \equiv p(z) = \det(zI - T) \qquad (6.6.45)$$

is the *Cayley-Hamilton* polynomial and $\lambda \in p^{-1}(0)$ is an eigenvalue then we can write

$$p = q \cdot r \ with \ q = (z - \lambda)^{\ell} \ and \ q^{-1}(0) \cap r^{-1}(0) = \emptyset \,, \qquad (6.6.46)$$

and hence

$$(\lambda I - T)^{-1}(0) \subseteq q^{-1}(0) \subseteq r(T)\mathbb{C}^{k} \ : \qquad (6.6.47)$$

all the eigenvectors $x \in (\lambda I - T)^{-1}(0)$ will be in the "column space" of the matrix $r(T)$.

References

1. Aiena, P.: Multipliers, Local Spectral Theory and Fredholm Theory. Kluwer, Boston (2004)
2. Aiena, P., Gonzalez, M.: Essentially incomparable Banach spaces and Fredholm theory. Proc. Royal Irish Acad. **93A**, 49–59 (1993)
3. Aiena, P., Aponte, E., Balzan, E.: Weyl type theorems for left and right polaroid operators. Int. Equ. Oper. Theory **66**, 1–20 (2010)
4. Allan, G.R.: On one-sided inverses in Banach algebras of holomorphic vector-valued functions. J. Lond. Math. Soc. **42**, 463–470 (1967)
5. Anderson, J.: On normal derivations. Proc. Amer. Math. Soc. **38**, 136–140 (1973)
6. Anderson, R.F.V.: The Weyl functional calculus. J. Funct. Anal. **4**, 240 267 (1969)
7. Anderson, J., Foias, C.: Properties which normal operators share with normal derivations and related properties. Pac. J. Math. **61**, 133–325 (1975)
8. Ara, P., Pedersen, G., Perera, F.: An infinite analogue of rings with stable rank one. J. Algebra **230**, 608–655 (2000)
9. Ara, P., Pedersen, G., Perera, F.: A closure operation in rings. Int. J. Math. **12**, 791–812 (2001)
10. Arias, M.L., Mbekhta, M.: On partial isometries in C*-algebras. Stud. Math. **205**, 71–82 (2011)
11. Arizmendi, H., Harte, R.E.: Almost open mappings in topological vector spaces. Proc. Royal Irish Acad. **99A**, 57–65 (1999)
12. Aupetit, B.: A Primer on Spectral Theory. Springer, Berlin (1991)
13. Aupetit, B., Mouton, H.T.: Trace and determinant in Banach algebras. Stud. Math. **121**, 115–136 (1996)
14. Baklouti, H.: T-Fredholm analysis and applications to operator theory. J. Math. Anal. Appl. **369**, 283–289 (2010)
15. Barnes, B.A.: Operator properties of module maps. J. Oper. Theory **33**, 79–104 (1995)
16. Barnes, B.A.: Common properties of the linear operators RS and SR. Proc. Amer. Math. Soc. **126**, 1055–1061 (1998)
17. Barnes, B.A.: The commutant of an abstract backward shift. Canad. Math. Bull. **43**, 21–24 (2000)
18. Barnes, B.A.: Spectral and Fredholm theory involving the diagonal of a bounded linear operator. Acta Sci. Math. **73**, 237–250 (2007)
19. Barnes, B.A., Murphy, G.J., Smyth, M.R.F., West, T.T.: Riesz and Fredholm theory in Banach algebras. Research Notes in Mathematics, vol. 67, Pitman London (1982)
20. Bermuda, T., Gonzalez, M.: Almost regular operators are regular. Bull. Korean Math. Soc. **38**, 205–210 (2001)
21. Bernau, S.J.: The spectral theorem for normal operators. J. Lond. Math. Soc. **40**, 478–486 (1965)
22. Bernau, S.J., Smithies, F.: A note on normal operators. Proc. Cam. Phil. Soc. **59**, 727–729 (1963)

© The Author(s), under exclusive license to Springer Nature Switzerland AG 2023
R. Harte, *Spectral Mapping Theorems*,
https://doi.org/10.1007/978-3-031-13917-8

23. Bhatia, R., Rosenthal, P.: How and why to solve the operator equation $AX - XB = Y$. Bull. Lond. Math. Soc. **29n**, 1–21 (1997)
24. Blackadar, B.: K-Theory for Operator Algebras. Springer, Berlin (1986)
25. Blackadar, B.: Operator Algebras. Springer, Berlin (2006)
26. Boasso, E.: On the Moore-Penrose inverse in C*-algebras. Extracts Math. **21**, 93–106 (2006)
27. Boasso, E.: Drazin spectra of Banach space operators and Banach algebra elements. J. Math. Anal. Appl. **359**, 48–55 (2009)
28. Boasso, E., Larotonda, A.: A spectral theory for solvable Lie algebras of operators. Pac. J. Math. **158**, 15–22 (1993)
29. Bonsall, F.F., Duncan, J.: Numerical Ranges I, II. Cambridge University Press, Cambridge (1971/1972)
30. Bonsall, F.F., Duncan, J.: Complete normed algebras. Springer, Berlin (1973)
31. Brits, R.M.: A correction to "Adjugates in Banach algebras". Proc. Amer. Math. Soc. **8**, 3021–3024 (2010)
32. Brits, R.M., Raubenheimer, H.: Finite spectra and quasinilpotent equivalence in Banach algebras. Czechoslovak Math. J. **62**, 1101–1116 (2012)
33. Brits, R.M., Lindeboom, L., Raubenheimer, H.: On ideals of generalized invertible elements in Banach algebras. Math. Proc. Royal Irish Acad. **105A**, 1–10 (2005)
34. Brown, A., Pearcy, C.: Spectra of tensor products of operators. Proc. Amer. Math. Soc. **17**, 162–166 (1966)
35. Bunce, J.W.: The joint spectrum of commuting non-normal operators. Proc. Amer. Math. Soc. **29**, 499–505 (1971)
36. Buoni, J.J., Harte, R.E., Wickstead, A.: Upper and lower Fredholm spectra. Proc. Amer. Math. Soc. **66**, 309–314 (1977)
37. Burlando, L., Harte, R.E.: On the closure of the invertibles. Pan Amer. J. **1**, 89–90 (1991)
38. Burlando, L., Harte, R.E.: On the closure of the invertibles in a von Neumann algebra. Colloq. Math. **69**, 157–165 (1995)
39. Caradus, S.R.: Operator Theory of the Pseudo Inverse. Queen's papers Pure Applied Mathematics, vol. 38. Queen's University Kingston, Kingston (1974)
40. Caradus, S.R.: Generalized Inverses and Operator Theory. Queen's Papers Pure Applied Mathematics, vol. 50. Queen's University Kingston, Kingston (1978)
41. Caradus, S.R., Pfaffenberger, W.E., Yood, B.: Calkin algebras of operators on Banach spaces. Dekker, New York (1974)
42. Carillo, A., Hernandez, C.: Spectra of constructs of a system of operators. Proc. Amer. Math. Soc. **91**, 426–432 (1984)
43. Cho, M., Takaguchi, M.: Boundary of Taylor's joint spectrum for two commuting operators. Sci. Rep. Horosaki Univ. **28**, 1–4 (1981)
44. Cho, M., Harte, R.E., Müller, V.: Transfinite ranges and the local spectrum. J. Math. Anal. Appl. **305**, 403–408 (2013)
45. Choi, M.-D., Davis, C.: The spectral mapping theorem for joint approximate spectrum. Bull. Amer. Math. Soc. **80**, 317–321 (1974)
46. Choukri, E., Illoussamen, E.H., Runde, V.: Gelfand theory for non commutative Banach algebras. Quart. J. Math. **53**, 161–172 (2002)
47. Coburn, L.A., Schechter, M.: Joint spectra and interpolation of oprators. J. Funct. Anal. **2**, 226–237 (1968)
48. Corach, G., Suarez, F.D.: Extensions of characters in commutative Banach algebras. Stud. Math. **85**, 199–202 (1987)
49. Corach, G., Duggal, B.P., Harte, R.E.: Extensions of Jacobson's lemma. Commun. Algebra **41**, 520–531 (2013)
50. Crimmins, T., Rosenthal, P.: On the decomposition of invariant subspaces. Bull. Amer. Math. Soc. **73**, 97–99 (1967)
51. Crownover, R.: Commutants of shifts on Banach spaces. Michican Math. J. **19**, 233–247 (1972)

52. Curto, R.E.: Fredholm and invertible n-tuples of operators. Trans. Amer. Math. Soc. **206**, 129–159 (1981)
53. Curto, R.E.: Spectral permanence for joint spectra. Trans. Amer. Msth. Soc. **270**, 659–665 (1982)
54. Curto, R.E.: Connections between Harte and Taylor spectrum. Rev. Roumaine Math. Pures Appl. **31**, 203–215 (1986)
55. Curto, R.E., Dash, A.T.: Browder spectral systems. Proc. Amer. Math. Soc. **103**, 407–413 (1988)
56. Curto, R.E., Fialkow, L.: The spectral picture of (L_A, R_B). J. Funct. Anal. **71**, 371–392 (1987)
57. Cvetkovic-Ilic, D., Harte, R.E.: Reverse order laws in C*-algebras. Linear Alg. Appl. **434**, 1388–1394 (2011)
58. Dash, A.T., Schechter, M.: Tensor products and joint spectra. Isr. J. Math. **8**, 191–193 (1970)
59. Davis, C., Rosenthal, P.: Solving linear operator equations. Canad. Math. J. **26**, 1384–1389 (1974)
60. Defant, A., Floret, K.: Tensor Norms and Operator Ideals. North-Holland, Amsterdam (1993)
61. Dineen, S., Harte, R.E.: Banach-valued axiomatic spectra. Stud. Math. **175**, 213–232 (2006)
62. Dineen, S., Mackey, M.: Confined Banach spaces. Arch. Math. **87**, 227–232 (2006)
63. Dineen, S., Harte, R.E., Taylor, C.: Spectra of tensor products I, basic theory. Math. Proc. Royal Irish Acad. **101A**, 177–196 (2001)
64. Dineen, S., Harte, R.E., Taylor, C.: Spectra of tensor products II, polynomial extensions. Math. Proc. Royal Irish Acad. **101A**, 197–220 (2001)
65. Dineen, S., Harte, R.E., Taylor, C.: Spectra of tensor products III, holomorphic properties. Math. Proc. Royal Irish Acad. **103A**, 61–92 (2003)
66. Dineen, S., Harte, R.E., Rivera, M.J.: An infinite dimensional functional calculus in Banach algebras with a modulus. Math. Nachr. **281**, 171–180 (2008)
67. Djordjevic, S.V., Kantun-Montiel, G.: On generalised T-Fredholm elements in Banach algebra. Math. Proc. Royal Irish Acad. **109A**, 61–66 (2009)
68. Djordjevic, S.V., Harte, R.E., Larson, D.R.: Partially hyper invariant subspaces. Oper. Matrices **6**, 97–106 (2012)
69. Djordjevic, D.S., Zivkovic-Zlatanovic, S., Harte, R.E.: On simple permanence. Quaest. Math. **38**, 515–528 (2015)
70. Douglas, R.G.: On majorization, factorization and range inclusion of operators. Proc. Amer. Math. Soc. **17**, 413–415 (1966)
71. Douglas, R.G.: Banach Algebra Techniques in Operator Theory. Academic Press, Cambridge (1972)
72. Dowson, H.R.: Spectral Theory of Linear Operators. Academic Press, Cambridge (1978)
73. Doxiadis, A.: Uncle Petros and Goldbach's Conjecture. Bloomsbury, London (1992)
74. Duggal, B.P., Harte, R.E.: Range-kernel orthogonality and range closure of an elementary operator. Monatsh. Math. **143**, 179–187 (2004)
75. Duggal, B.P., Jeon, I.H., Harte, R.E.: Polaroid operators and Weyl's theorem. Proc. Amer. Math. Soc. **132**, 1345–1349 (2004)
76. Edgar, G., Ernest, J., Lee, S.G.: Weighing operator spectra. Ind. Math. J. **21**, 61–80 (1971)
77. Effros, E.G., Ruan, Z.-J.: On the abstract characterization of operator spaces. Proc. Amer. Math. Soc. **119**, 579–584 (1993)
78. Effros, E.G., Ruan, Z.-J.: Operator Spaces. London Mathematical Society Monographs. Clarendon, Oxford (2001)
79. Embry, M.R.: A connection between commutivity and separation of spectra of linear operators. Acta Sci. Math. **32**, 235–237 (1971)
80. Embry, M.R.: Factorization of operators on Banach spaces. Proc. Amer. Math. Soc. **38**, 587–590 (1973)
81. Embry, M.R., Rosenblum, M.: Spectra, tensor products and linear operator equations. Pac. J. Math. **53**, 95–107 (1974)
82. Ernest, J.: Charting the Operator Terrain. Memoirs of the American Mathematical Society, vol. 171. American Mathematical Society, Providence (1976)

83. Fainstein, A.S.: Taylor spectrum for families of operators generating nilpotent Lie algebras. J. Oper. Theory **29**, 3–27 (1993)
84. Fialkow, L.: A note on quasisimilarity of operators. Acta Sci. Math. **30**, 67–85 (1977)
85. Fialkow, L.: Spectral properties of elementary operators. Acta Sci. Math. **46**, 269–282 (1983)
86. Fialkow, L.: Spectral properties of elementary operators II. Trans. Amer. Math. Soc. **290**, 415–429 (1985)
87. Finch, J.K.: The single-valued extension property on a Banach space. Pac. J. Math. **58**, 61–69 (1958)
88. Fong, C.-K.: Normal operators on Banach spaces. Glasgow Math. J. **20**, 163–168 (1979)
89. Fong, C.-K., Sourour, A.R.: On the operator identity $\sum A_k X B_k \equiv 0$. Canad. J. Math. **31**, 845–857 (1979)
90. Garimella, R.V., Hrynkiv, V., Sourour, A.R.: An operator equation, KdV equation and invariant subspaces. Proc. Amer. Math. Soc. **138**, 717–724 (2010)
91. George, A., Ikramov, K.D.: Common invariant subspaces of two matrices. Linear Alg. Appl. **287**, 171–179 (1999)
92. Ghez, P., Lima, R., Roberts, J.E.: W* categories. Pac. J. Math. **120**, 79–109 (1985)
93. Goldberg, S.: Unbounded Linear Operators. McGraw-Hill, New York (1966)
94. Gonzalez, M.: A perturbation result for generalized Fredholm operators on the boundary of the invertible group. Proc. Royal Irish Acad. **86A**, 123–126 (1986)
95. Gonzalez, M.: Null spaces and ranges of polynomials of operators. Pub. Mat. Univ. Barcelona **32**, 167–170 (1988)
96. Gonzalez, M.: Essentially incomparable Banach spaces. Math. Zeit **215**, 621–629 (1994)
97. Gonzalez, M., Harte, R.E.: The death of an index theorem. Proc. Amer. Math. Soc. **108**, 151–156 (1990)
98. Goodearl, K.R.: Notes on Real and Complex C* Algebras. Shiva Publishing, Delhi (1982)
99. Grabner, S.: Ascent, descent, and compact perturbations. Proc. Amer. Math. Soc. **71**, 79–80 (1978)
100. Grabner, S.: Uniform ascent and descent of bounded operators. J. Math. Soc. Japan **34**, 317–337 (1982)
101. Gramsch, B., Lay, D.C.: Spectral mapping theorems for essential spectra. Math. Ann. **192**, 17–32 (1971)
102. Grobler, J.J., Raubenheimer, H.: Spectral properties of elements in different Banach algebras. Glasgow Math. J. **33**, 11–20 (1991)
103. Groenewald, L., Harte, R.E., Raubenheimer, H.: Perturbation by inessential and Riesz elements. Quaest. Math. **12**, 439–446 (1989)
104. Hadwin, D.W.: An operator-valued spectrum. Ind. Univ. Math. J. **26**, 329–340 (1977)
105. Hadwin, D.W., Nordgren, E., Rosenthal, P.: On the operator equation $AXB + CYD = Z$. Oper. Matrices **1**(2), 199–207 (2007)
106. Han, D., Larson, D.R., Pan, Z., Wogen, W.: Extensions of operators. Ind. Univ. Math. J. **53**, 1151–1169 (2004)
107. Harte, R.E.: Modules over a Banach algebra. Doctoral Disseration Cambridge (1965)
108. Harte, R.E.: A generalization of the Hahn-Banach theorem. J. Lond. Math. Soc. **40**, 283–287 (1965)
109. Harte, R.E.: A theorem of isomorphism. Proc. Lond. Math. Soc. **16**, 753–765 (1966)
110. Harte, R.E.: The spectral mapping theorem in several variables. Bull. Amer. Math. Soc. **78**, 871–875 (1972)
111. Harte, R.E.: Spectral mapping theorems. Proc. Royal Irish Acad. **72**(A), 89–107 (1972)
112. Harte, R.E.: Relatively invariant systems and the spectral mapping theorem. Bull. Amer. Math. Soc. **79**, 138–142 (1973)
113. Harte, R.E.: The spectral mapping theorem for quasicommuting systems. Proc. Royal Irish Acad. **73**(A), 7–18 (1973)
114. Harte, R.E.: Spectral mapping theorems on a tensor product. Bull. Amer. Math. Soc. **79** 367–372 (1973)

115. Harte, R.E.: Tensor products, multiplication operators and the spectral mapping theorem. Proc. Royal Irish Acad. **73(A)**, 285–302 (1973)
116. Harte, R.E.: The spectral mapping theorem in many variables. Proceedings of Symposium "Uniform Algebras", pp. 59–63. University of Aberdeen, Aberdeen (1973)
117. Harte, R.E.: Commutivity and separation of spectra II. Proc. Royal Irish Acad. **74(A)**, 239–244 (1974)
118. Harte, R.E.: A Silov boundary for systems? In: Proceedings of Symposium "Algebras in Analysis", University of Birmingham 1973, pp. 268–271. Academic Press, Cambridge (1975)
119. Harte, R.E.: A problem of mixed interpolation. Math. Zeit. **143**, 149–153 (1975)
120. Harte, R.E.: The exponential spectrum in Banach algebras. Proc. Amer. Math. Soc. **58**, 114–118 (1976)
121. Harte, R.E., Wickstead, A.: Upper and lower Fredholm spectra II. Math. Zeit. **154**, 253–256 (1977)
122. Harte, R.E.: Berberian-Quigley and the ghost of a spectral mapping theorem. Proc. Royal Irish Acad. **78(A)**, 63–68 (1978)
123. Harte, R.E.: Invertibility, singularity and Joseph L. Taylor. Proc. Royal Irish Acad. **81(A)**, 71–79 (1981)
124. Harte, R.E.: Fredholm theory relative to a Banach algebra homomorphism. Math. Zeit. **179**, 431–436 (1982)
125. Harte, R.E.: A quantitative Schauder theorem. Math. Zeit. **185**, 243–245 (1984)
126. Harte, R.E.: Almost open mappings between normed spaces. Proc. Amer. Math. Soc. **90**, 243–249 (1984)
127. Harte, R.E.: Spectral projections. Irish Math. Soc. Newslett. **14**, 10–15 (1984)
128. Harte, R.E.: A matrix joke. Irish Math. Soc. Newslett. **14**, 13–16 (1985)
129. Harte, R.E.: Fredholm, Weyl and Browder theory. Proc. Royal Irish Acad. **85(A)**, 151–176 (1986)
130. Harte, R.E.: Almost exactness in normed spaces. Proc. Amer. Math. Soc. **100**, 257–265 (1987)
131. Harte, R.E.: Regular boundary elements. Proc. Amer. Math. Soc. **99**, 328–330 (1987)
132. Harte, R.E.: Invertibility and Singularity for Bounded Linear Operators. Monographs and Textbooks in Pure Mathematics, vol. 109. Marcel Dekker, New York (1988)
133. Harte, R.E.: Invertibility and singularity. Marcel Dekker, New York (1988)
134. Harte, R.E.: A note on generalized inverse functions. Proc. Amer. Math. Soc. **104**, 551–552 (1988)
135. Harte, R.E.: The ghost of an index theorem. Proc. Amer. Math. Soc. **106**, 1031–1034 (1989)
136. Harte, R.E.: Cayley-Hamilton for eigenvalues. Bull. Irish Math. Soc. **22**, 66–68 (1989)
137. Harte, R.E.: Compound matrices revisited (determinants the hard way). Linear Alg. Appl. **117**, 156–159 (1989)
138. Harte, R.E.: Invertibility and singularity for operator matrices. Proc. Royal Irish Acad. **88(A)**, 103–118 (1989)
139. Harte, R.E.: On quasinilpotents in rings. Pan Amer. J. **1**, 10–16 (1991)
140. Harte, R.E., Mathieu, M.: Enlargements of almost open mappings. Proc. Amer. Math. Soc. **96**, 247–248 (1986)
141. Harte, R.E.: Fredholm, Weyl and Browder theory II. Proc. Royal Irish Acad. **91A**, 79–88 (1991)
142. Harte, R.E.: On the punctured neighbourhood theorem. Math. Zeit. **207**, 391–394 (1991)
143. Harte, R.E.: Eine kleine gapmusik. Pan Amer. J. **2**, 101–102 (1992)
144. Harte, R.E.: Polar decomposition and the Moore-Penrose inverse. Pan Amer. J. **2**, 71–76 (1992)
145. Harte, R.E.: Taylor exactness and Kaplansky's lemma. J. Oper. Theory **25**, 399–416 (1992)
146. Harte, R.E.: Taylor exactness and Kato's jump. Proc. Amer. Math. Soc. **119**, 793–802 (1993)
147. Harte, R.E.: On rank one elements. Stud. Math. **117**, 73–77 (1995)
148. Harte, R.E.: On Kato non singularity. Stud. Math. **117**, 107–114 (1996)
149. Harte, R.E.: Exactness plus skew exactness equals invertibility. Proc. Royal Irish Acad. **97A**, 15–18 (1997)

150. Harte, R.E.: On criss-cross commutivity. J. Oper. Theory **37**, 303–309 (1997)
151. Harte, R.E.: Unspectral sets. Rendiconti del Circulo Matematico di Palermo **56**, 69–77 (1997)
152. Harte, R.E.: Almost regularity III. Math. Proc. Royal Irish Acad. **99A**, 155–162 (1999)
153. Harte, R.E.: Spectral theory in many variables. In: Proceedings of TGRC-KOSEF Kyungpook National University Korea (2000)
154. Harte, R.E.: Criss-cross commutivity II. J. Oper. Theory **46**, 39–43 (2001)
155. Harte, R.E.: Block diagonalization in Banach algebras. Proc. Amer. Math. Soc. **129**, 181–190 (2001)
156. Harte, R.E.: Variations on a theme of Kaplansky. Filomat **16**, 19–30 (2002)
157. Harte, R.E.: Arens-Royden and the spectral landscape. Filomat **16**, 31–42 (2002)
158. Harte, R.E.: On spectral boundedness. J. Korean Math. Soc. **40**, 307–317 (2003)
159. Harte R.E.: Skew exactness and range-kernel orthogonality. Filomat (Nis) **19**, 19–33 (2005)
160. Harte, R.E.: The triangle inequality in C* algebras. Filomat (Nis) **20**, 51–53 (2006)
161. Harte, R.E.: Tim Starr, Trevor West and a positive contribution to operator theory. Filomat (Nis) **21**, 129–135 (2007)
162. Harte, R.E.: Skew exactness and range-kernel orthogonality II. J. Math. Anal. Appl. **347**, 370–374 (2008)
163. Harte, R.E.: On local spectral theory. Oper. Theory Adv. Appl. **187**, 175–183 (2008)
164. Harte, R.E.: Hermitian subspaces and Fuglede operators. Funct. Anal. Approx. Comput. **2**, 19–32 (2010)
165. Harte, R.E.: On local spectral theory II. Funct. Anal. Approx. Comput. **2**, 67–71 (2010)
166. Harte, R.E.: Spectral permanence. Irish Math. Soc. Bull. **69**, 33–46 (2012)
167. Harte, R.E.: Non commutative Müller regularity. Funct. Anal. Approx. Comput. **6**, 1–7 (2014)
168. Harte, R.E.: Spectral permanence II. Funct. Anal. Aprox. Comput. **7**, 1–7 (2015)
169. Harte, R.E.: Residual quotients. Funct. Anal. Approx. Comput. **7**, 67–74 (2015)
170. Harte, R.E.: On non commutative Taylor invertibility. Oper. Math. **10**, 1117–1131 (2016)
171. Harte, R.E.: Spectral disjointness and the Euclidean algorithm. Math. Proc. Royal Irish Acad. **118A**, 65–69 (2018)
172. Harte, R.E.: Spectral disjointness and invariant subspaces. Maltepe J. Math. **1**, 56–65 (2019)
173. Harte, R.E.: On spectral number theory. Funct. Anal. Approx. Comput. **13**, 1–6 (2021)
174. Harte, R.E., Cvetkovic-Ilic, D.: The algebraic closure in rings. Proc. Amer. Math. Soc. **135**, 3547–3552 (2007)
175. Harte, R.E., Cvetkovic-Ilic, D.: On Jacobson's lemma and Drazin invertibility. Appl. Math. Lett. **23**, 417–420 (2010)
176. Harte, R.E., Cvetkovic-Ilic D.: The spectral topology in rings. Stud. Math. **200**, 267–278 (2010)
177. Harte, R.E., Hernandez, C.: On the Taylor spectrum of left-right multipliers. Proc. Amer. Math. Soc. **126**, 103–118 (1998)
178. Harte, R.E., Hernandez, C.: Adjugates in Banach algebras. Proc. Amer. Math. Soc. **134** 1397–1404 (2005)
179. Harte, R.E., Hernandez C.: Adjugates on linear algebras, Funct. Amal. Approx. Comp. **14** 23–39 (2022)
180. Harte, R.E., Keogh, G.: Touché Rouché. Bull. Korean Math. Soc. **40**, 215–221 (2003)
181. Harte, R.E., Kim, A.-H.: Weyl's theorem, tensor products and multiplication operators. J. Math. Anal. Appl. **336**, 1124–1131 (2007)
182. Harte, R.E., Larson, D.R.: Skew exactness perturbation. Proc. Amer. Math. Soc. **132**, 2603 (2004)
183. Harte, R.E., Lee, W.Y.: The punctured neighbourhood theorem for incomplete spaces. J. Oper. Theory **30**, 217–226 (1993)
184. Harte, R.E., Lee, W.Y.: An index formula for chains. Stud. Math. **116**, 283–294 (1995)
185. Harte, R.E., Lee, W.Y.: A note on the punctured neighbourhood theorem. Glasgow Math. J. **39**, 269–273 (1997)
186. Harte, R.E., Lee, W.Y.: Another note on Weyl's theorem. Trans. Amer. Math. Soc. **349**, 2115–2124 (1997)

187. Harte, R.E., Lee, W.Y.: On the bounded closure of the range of an operator. Proc. Amer. Math. Soc. **125**, 2313–2318 (1997)
188. Harte, R.E., Mbekhta, M.: Generalized inverses in C*-algebras. Stud. Math. **103**, 71–77 (1992)
189. Harte, R.E., Mbekhta, M.: Almost exactness in normed spaces II. Stud. Math. **117**, 101–105 (1996)
190. Harte, R.E., Mbekhta, M.: Generalized inverses in C*-algebras II. Stud. Math. **106**, 129–138 (1993)
191. Harte, R.E., Mouton, S.: Linking the boundary and the exponential spectrum via the restricted spectrum. J. Math. Anal. Appl. **454**, 730–745 (2017)
192. Harte, R.E., O'Searcoid, M.: Positive elements and the B* condition. Math. Zeit. **193**, 1–9 (1986)
193. Harte, R.E., Shannon, G.P.: Closed subspaces of operator ranges. Proc. Amer. Math. Soc. **118**, 171–173 (1993)
194. Harte, R.E., Stack, C.M.: Invertibility of spectral triangles. Oper. Matrices **1**, 445–453 (2007)
195. Harte, R.E., Stack, C.M.: Separation of spectra for block triangles. Proc. Amer. Math. Soc. **136**, 3159–3162 (2008)
196. Harte R.E., Raubenheimer, H.: Fredholm, Weyl and Browder theory III. Proc. Royal Irish Acad. **95A**, 11–16 (1995)
197. Harte, R.E., Stack, C.M.: On left-right consistency in rings II. Math. Proc. Royal Irish Acad. **106**, 219–223 (2006)
198. Harte, R.E., Taylor, C.: On vector-valued spectra. Glasgow Math. J. **42**, 247–253 (2000)
199. Harte, R.E., Taylor, C.: The functional calculus and all that. Rendiconti del Circulo Matematico di Palermo **73**, 77–89 (2004)
200. Harte, R.E., Wickstead, A.: Boundaries, hulls and spectral mapping theorems. Proc. Royal Irish Acad. **81(A)**, 201–208 (1981)
201. Harte, R.E., Lee, W.Y., Littlejohn, L: On generalized Riesz points. J. Oper. Theory **47**, 187–196 (2002)
202. Harte, R.E., Kim, Y.O., Lee, W.Y.: Spectral pictures of AB and BA. Proc. Amer. Math. Soc. **134**, 105–110 (2005)
203. Harte, R.E., Djordjevic, D.S., Stack, C.M.: On left-right consistency in rings. Math. Proc. Royal Irish Acad. **106**, 11–17 (2006)
204. Harte, R.E., Djordjevic, D.S., Stack, C.M.: On left-right consistency in rings III. Quaest. Mat. **34**, 335–339 (2011)
205. Harte, R.E., Hernandez, C., Stack, C.M.: Exactness and the Jordan form. Funct. Anal. Approx. Comput. **3**, 1–7 (2011)
206. Hartwig, R.: Block generalized inverses. Arch Rat. Mech. Anal. **61**, 197–251 (1976)
207. Ichinose, T.: Spectral properties of linear operators I. Trans. Amer. Math. Soc. **235**, 75–113 (1978)
208. Kantun-Montiel, G., Djordjevic, S.V., Harte, R.E.: On semigroup inverses. Funct. Anal. Approx. Comput. **1**, 11–18 (2009)
209. Kato, T.: Perturbation Theory for Linear Operators. Springer, Berlin (1966)
210. Kim, A.-H., Duggal, B.P., Harte, R.E.: Weyl's theorem, tensor products and multiplication operators II. Glasgow Math. J. **52**, 703–709 (2010)
211. Kisil, V.V.: Möbius transformations and the monogenic functional calculus. Electron. Res. Announc. Amer. Math. Soc. **2**, 26–33 (1996)
212. Kisil, V.V.: Spectrum as support of functional calculus. Leeds Unuiversity Preprint math/0208249
213. Kitson, D., Harte, R.E.: On Browder tuples. Acta Sci. Math. (Szeged) **75**, 665–677 (2009)
214. Kitson, D., Harte, R.E., Hernandez, C.: Weyl's theorem and tensor products: a counterexample. J. Math. Anal. Appl. **378**, 128–132 (2011)
215. Koliha, J.J.: A generalized Drazin inverse. Glasgow Math. J. **38**, 367–381 (1996)
216. Koliha, J.J.: Isolated spectral points. Proc. Amer. Math. Soc. **124**, 3417–3424 (1996)
217. Koliha, J.J., and Poon, P.W., Spectral sets, II Rendiconti Mat. **47**, 293–310, 1998

218. Kordula, V., Müller, V.: On the axiomatic theory of spectrum. Stud. Math. **119**, 109–208 (1996)
219. Koszul, J.L.: Homologie et cohomologie de algebres de Lie. Bull. Soc. Math. France **78**, 65–127 (1950)
220. Kovarik, Z.V.: Similarity and interpolation between projectors. Acta Sci. Math. (Szeged) **39**, 341–351 (1977)
221. Krishnan, E., Nambooripad, K.S.S.: The semigroup of Fredholm operators. Forum Math. **5**, 313–368 (1993)
222. Kuiper, N.: The homotopy type of the unitary group of Hilbert space. Topology **3**, 19–30 (1965)
223. Labrousse, J.P.: Les opérateurs quasi-Fredholm. Rend. Circ. Mat. Palermo **29**, 161–258 (1980)
224. Laffey, T.J., West, T.T.: Fredholm commutators. Proc. Royal Irish Acad. **82A**, 129–140 (1982)
225. Lance, E.C.: C* Modules. London Mathematical Society Lecture Note. Cambridge University Press, Cambridge (1995)
226. Laursen, K.B., Neumann, M.M.: Introduction to Local Spectral Theory. Clarendon Press, Oxford (2000)
227. Lay, D.C.: Spectral analysis using ascent, descent, nullity and defect. Math. Ann. **184**, 197–214 (1970)
228. Lebow, L., Schechter, M., Lee, S.-G., Vu, Q.-Ph.: Semigroups of operators and measures of non compactness. J. Funct. Anal. **7**, 1–26 (1971)
229. Lee, S.-G., Vu, Q.-Ph.: Simultaneous solutions of operator Sylvester equations. Studia Math. **222**, 87–96 (2014)
230. Littlewood, J.E.: A Mathematician's Miscellany. Methuen, London (1953)
231. Lomonosov, V.: Invariant subspaces for operators commuting with compact operators. Funct. Anal. Appl. **7**, 213–214 (1973)
232. Lubansky, R.A.: Koliha-Drazin invertibles form a regularity. Math. Proc. Royal Irish Acad. **107A**, 137–141 (2007)
233. Lumer, G., Rosenblum, M.: Linear operator equations. Proc. Amer. Math. Soc. **10**, 32–41 (1959)
234. Martinez-Melendez, A., Wawrzynczyk, A.: An approach to joint spectra. Ann. Polon. Math. **72**, 131–144 (2000)
235. Martinez-Avendano, R.: Eigenmatrices and operators commuting with finite-rank operators. Linear Alg. Appl. 419, 739–749 (2006)
236. Mary, X.: On the converse of a theorem of Harte and Mbekhta. Stud. Math. **184**, 149–152 (2008)
237. Mary, X.: On generalized inverses and Green's relations. Linear Alg. Appl. **434**, 1830–1844 (2011)
238. Mary, X., Patricio, P.: The inverse along a lower triangular matrix. Appl. Math. Comput. **219**, 886–891 (2012)
239. Mascerenhas, H., Santos P., Seidel, M.: Quasi-banded operators, convolutions with almost periodic or nquasi-continuous data, and thrir approximations. J. Math. Anal. Appl. **418**, 938–963 (2014)
240. Mathieu, M.: Elementary operators on prime C*-algebras I. Math. Ann. **284**, 223–244 (1989)
241. Mattila, K.: Normal operators and proper boundary points of the spectra of operators on a Banach space. Ann. Acad. Sci. Fenn. **19**, 48pp. (1978)
242. Mattila, K.: Complex strict and uniform convexity and hyponormal operators. Mth. Proc. Cam. Phil. Soc. **96**, 483–493 (1984)
243. Mattila, K.: A class of hyponormal operators and weak* continuity of hermitian operators. Ark. Mat. **25**, 265–274 (1987)
244. Mbekhta, M.: Résolvent généralisé et théorie spectrale. J. Oper. Theory **21**, 69–105 (1989)
245. Mbekhta, M.: Partial isometries and generalized inverses. Acta Sci. Math. (Szeged) **70**, 767–781 (2004)

246. Mbekhta, M., Müller, V.: On the axiomatic theory of spectrum II. Stud. Math. **119**, 129–147 (1996)
247. McCoy, N.H.: On quasicommutative natrices. Trans. Amer. Math. Soc. **36**, 327–340 (1934)
248. McIntosh, A., Pryde, A., Ricker, W.: Comparison of joint spectra for certain classes of commuting operators. Stud. Math. **88**, 23–36 (1988)
249. Mouton, H.T.: On inessential ideals in Banach algebras. Quaest. Math. **17**, 59–66 (1994)
250. Mouton, S.: On the boundary spectrum in Banach algebras. Bull. Austral. Math. Soc. **74**, 230–246 (2006)
251. Mouton, S.: Mapping and continuity properties of the boundary spectrum in Banach algebras. Illinois J. Math. **53**, 757–767 (2009)
252. Mouton, S.: Generalized spectral perturbation and the boundary spectrum. Czech. Math. J. **71**, 603–621 (2021)
253. Mouton, H.T., Raubenheimer, H.: On rank one and finite elements in Banach algebras. Stud. Math. **104**, 212–219 (1993)
254. Mouton, S., Mouton, H.T., Raubenheimer, H.: Ruston elements and Fredholm theory relative to aebitrary homomorphisms. Quaest. Math. **34**, 341–359 (2011)
255. Müller, V.: Slodkowski spectra and higher Shilov boundaries. Stud. Math. **105**, 69–75 (1993)
256. Müller, V.: On the regular spectrum. J. Oper. Theory **31**, 363–380 (1994)
257. Müller, V.: The splitting spectrum differs from the Taylor spectrum. Stud. Math. **123**, 291–294 (1997)
258. Müller, V.: Axiomatic theory of the spectrum III - semiregularities. Stud. Math. **142**, 159–169 (2000)
259. Müller, V.: Spectral Theory of Linear Operators. Birkhäuser, Basel (2007)
260. Murphy, G.J.: C*-algebras and Operator Theory. Academic Press, Cambridge (1990)
261. Murphy, G.J.: The index group, the exponential spectrum, and some spectral containment theorems. Proc. Royal Irish Acad. **92A**, 229–238 (1992)
262. Murphy, G.J.: Fredholm index theory and the trace. Proc. Royal Irish Acad. **94A**, 161–166 (1994)
263. Nevanlinna, O.: Sylvester equations and polynomial separation of spectra. Oper. Math. **13**, 867–885 (2019)
264. Newburgh, J.D.: The variation of spectra. Duke Math. J. **18**, 165–176 (1951)
265. Northcott, D.G.: Ideal Theory. Cambridge Tracts in Mathematics, vol. 42. Cambridge University Press (1960)
266. O'Searcoid, M.: Elements of Abstract Analysis. Springer Undergraduate Mathematics Series, vol. 515. Springer, Berlin (2002)
267. Ott, C.: A note on a paper of Boasso and Larotonda, "A spectral theory for solvable Lie algebras...". Pac. J. Math. **173**, 173–179 (1996)
268. Palmer, T.W.: Characterization of C*-algebras. Bull. Amer. Math. Soc. **74**, 538–540 (1968)
269. Palmer, T.W.: Banach algebras and the general theory of *-algebras. Cambridge University Press, Cambridge (1994)
270. Radjavi, H., Rosenthal, P.: Invariant Suspaces. Springer, Berlin (1973)
271. Rakocevic, V.: Approximate point spectrum and commuting compact perturbations. Glasgow Math. J. **28**, 193–198 (1986)
272. Rakocevic, V.: Moore-Penrose inverse in Banach algebras. Proc. Royal Iriah Acad. **88A**, 57–60 (1988)
273. Rakocevic, V.: Generalized spectrum and commuting compact perturbations. Proc. Edinburgh Math. Soc. **36**, 197–209 (1993)
274. Rakocevic, V.: Semi-Fredholm operators with finite ascent or descent and perturbations. Proc. Amer. Math. Soc. **123**, 3823–3825 (1995)
275. Raubenheimer, H.: On quasinilpotent equivalence of finite rank elements in Banach algebras. Czech Math. J. **60**, 589–596 (2010)
276. Razpet, M.: The quasinilpotent equivalence in Banach algebras. J. Math. Anal. Appl. **166**, 378–385 (1992)
277. Read, C.J.: All primes have closed range. Bull. Lond. Math. Soc. **33**, 341–346 (2001)

278. Rieffel, M.: Dimension and stable rank in the K-theory of C*-algebras. Proc. Lond. Math. Soc. **46**, 301–333 (1983)
279. Rosenthal, D., Rosenthal, D., Rosenthal, P.: A Readable Introduction to Real Mathematics. Springer Undergraduate Texts in Mathematics. Springer, Berlin (2021)
280. Rynne, B.P.: Tensor products and Taylor's joint spectrum in Hilbert space. Proc. Royal Irish Acad. **88A**, 71–83 (1988)
281. Saphar, P.: Contribution a l'étude des applications linéares dans un espace de Banach. Bull. Soc. Math. de France **92**, 363–384 (1964)
282. Schechter, M., Snow, M.: The Fredholm spectrum of tensor products. Proc. Royal Irish Acad. **75A**, 121–128 (1975)
283. Schmoeger, C.: On a class of generalized Fredholm operators I-VII. Demonstratio Math. **33**, 30–33 (2017)
284. Schmoeger, C.: On the operator equation $ABA = A^2$ and $BAB = B^2$. Publ. Inst. Mat. Beograd **78**, 127–133 (2005)
285. Schmoeger, C.: Common spectral properties of linear operators such that $ABA = A^2$ and $BAB = B^2$. Publ. Inst. Mat. Beograd **79**, 109–114 (2006)
286. Slodkowski, Z.: An infinite family of joint spectra. Stud. Math. **61**, 139–255 (1977)
287. Slodkowski, Z., Zelazko, W.: On joint spectra of commuting families of operators. Stud. Math. **50**, 127–148 (1974)
288. Smyth, M.R.F.: A note on the action of an operator on its centraliser. Proc. Royal Irish Acad. **74A**, 297–298 (1974)
289. Smyth, M.R.F.: Riesz theory in Banach algebras. Math. Zeit. **145**, 144–155 (1975)
290. Smyth, M.R.F.: Riesz algebras. Proc. Royal Irish Acad. **76A**, 327–333 (1976)
291. Smyth, M.R.F., West, T.T.: The spectral radius formula in quotient algebras. Math. Zeit. **145**, 157–161 (1975)
292. Snow, M.: A joint Browder essential spectrum. Proc. Royal Irish Acad. **75A**, 129–131 (1975)
293. Song, Y.-H., Kim, A.-H.: Weyl's theorem for tensor products. Glasgow Math. J. **46**, 301–304 (2004)
294. Stampfli, J.G.: Compact perturbations, normal eigenvalues and a priblem of Salinas. J. Lond. Math. Soc. **9**, 165–175 (1974)
295. Taylor, A.E.: Theorems on ascent, descent, nullity snd defect of linear operators. Math. Ann. **163** 18–49 (1966)
296. Taylor, M.E.: Functions of aeveral self-adjoint operators. Proc. Amer. Math. Soc. **19**, 91–98 (1968)
297. Taylor, J.L.: A joint spectrum for several commuting operators. J. Funct. Anal. **6**, 172–191 (1970)
298. Taylor, J.L.: Banach algebras and topology. In: Proceedings of Symposium "Algebras in Analysis", University of Birmingham 1973, pp. 268–271. Academic Press, Cambridge (1975)
299. Treese, G.W., Kelly, E.P.: Generalized Fredholm operators and the boundary of the maximal group. Proc. Amer. Math. Soc. **67**, 123–128 (1977)
300. Vasilescu, F.-H.: A characterization of the joint spectrum in Hilbert spaces. Rev. Roum. Math. Pures Appl. **22**, 1003–1009 (1977)
301. Vasilescu, F.-H.: On pairs of commuting operators. Stud. Math. **62**, 203–207 (1978)
302. Vasilescu, F.-H.: Mathematical Reviews MR 95d 1587–1605
303. Voiculescu, D.: A non-commutative Weyl-von Neumann theorem. Rev. Roum. Math. Pures Appl. **21**, 91–113 (1976)
304. Waelbroeck, L.: Le calcul symbolique dans les algèbres commutatives. J. Math. Pures Appl. **33**, 147–186 (1954)
305. Waelbroeck, L.: Topological Vector Spaces and Algebras. Lecture Notes in Mathematics, vol. 230. Springer, Berlin (1973)
306. Wall, C.T.C.: A geometric introduction to topology. Addison-Wesley, Boston (1972)
307. Wawrzynczyk, A.: On ideals of topological zero divisors. Stud. Math. **142**, 245–251 (2000)
308. Wawrzynczyk, A.: Ideals of functions which achieve zero on a compact set. Bull. Soc. Mat. Mexicana **7**, 117–121 (2001)

309. Wawrzynczyk, A.: Harte's theorem for Waelbroeck algebras. Math. Proc. Royal Irish Acad. 105A, 71–77 (2005)
310. Wawrzynczyk, A.: Schur Lemma and the spectral mapping formula. Bull. Pol. Acad. Sci. **55**, 63–69 (2007)
311. Wawrzynczyk, A.: Joint spectra in Waelbroeck algebras. Bol. Soc. Math. Mexicana **13**, 321–343 (2017)
312. Wawrzynczyk, A.: Subsets of nonempty joint spectrum in topological algebras. Math. Bohemica **143**, 441–448 (2018)
313. West, T.T.: The decomposition of Riesz operators. Proc. Lond. Math. Soc. **16**, 737–752 (1966)
314. West, T.T.: A Riesz-Schauder theorem for semi-Fredholm operators. Proc. Royal Irish Acad. **87A**, 137–146 (1987)
315. Williams, D.P.: A (very) short course on C*-algebras, Preprint
316. Williams, J.P.: On the range of a derivation. Pac. J. Math. **35**, 273–279 (1971)
317. Wrobel, V.: The boundary of Taylor's joint spectrum for two commuting operators. Stud. Math. **84**, 105–111 (1986)
318. Xue, Y.F.: A note about a theorem of R. Harte. Filomat (Nis) **22**, 95–98 (2008)
319. Yang, K.W.: Index of Fredholm operators. Proc. Amer. Math. Soc. **41**, 329–330 (1973)
320. Zabreiko, P.P.: A theorem for subadditive functionals. Funct. Anal. Appl. **3**, 70–72 (1969)
321. Zame, W.R.: Existence uniqueness and continuity of functional calculus homomorphisms. Proc. Lond. Math. Soc. **39**, 73–92 (1979)
322. Zelazko, W.: A characterization of multiplicative linear functionals in complex Banach algebras. Stud. Math. **30**, 83–85 (1968)
323. Zelazko, W.: A characterization of Shilov boundary in function algebras. Comment. Math. **14**, 63–68 (1970)
324. Zelazko, W.: On a certain class of non-removable ideals in Banach algebras. Stud. Math. **44**, 87–92 (1972)
325. Zelazko, W.: Axiomatic approach to joint spectra I. Stud. Math. **64**, 249–261 (1979)
326. Zemanek, J.: The stability radius of a semi-Fredholm operator. Int. Eq. Oper. Theory **8**, 137–144 (1985)
327. Zeng, Q., Zhong, H.: Common properties of bounded operators AC and BA: spectral theory. Math. Nachr. **287**, 717–725 (2014)
328. Zivkovic-Zlatanovic, S., Djordjevic, D.S., Harte, R.E.: On left and right Browder operators. J. Korean Math. Soc. **48**, 1053–1063 (2011)
329. Zivkovic-Zlatanovic, S., Djordjevic, D.S., Harte, R.E.: Spectral permanence and the Moore-Penrose inverse. Proc. Amer. Math. Soc. **140**, 8237–8245 (2012)
330. Zivkovic-Zlatanovic, S., Djordjevic, D.S., Harte, R.E.: Ruston, Riesz and perturbation classes. J. Math. Anal. Appl. **389**, 871–886 (2012)
331. Zivkovic-Zlatanovic, S., Djordjevic, D.S., Harte, R.E.: Polynomially Riesz perturbations. J. Math. Anal. Appl. **408**, 442–451 (2013)

Index

© The Author(s), under exclusive license to Springer Nature Switzerland AG 2023
R. Harte, *Spectral Mapping Theorems*,
https://doi.org/10.1007/978-3-031-13917-8

Symbol Index

© The Author(s), under exclusive license to Springer Nature Switzerland AG 2023
R. Harte, *Spectral Mapping Theorems*,
https://doi.org/10.1007/978-3-031-13917-8

185

Printed in the United States
by Baker & Taylor Publisher Services

Printed in the United States
by Baker & Taylor Publisher Services